111 range and forage plants of the Canadian prairies

J. Looman
Research Station
Swift Current, Saskatchewan

Research Branch
Agriculture Canada
Publication 1751
1983

Replaces Publication 964, "Ninety-nine range forage plants of the Canadian prairies",
by J. B. Campbell, K. F. Best, and A. C. Budd

Available in Canada through
your local bookseller

or by mail from
Canada Communication Group — Publishing
Ottawa, Canada K1A 0S9

Catalogue No. A53-1751-1983E
ISBN 0-660-11387-2

Canadian Cataloguing in Publication Data

Looman, J.

111 range and forage plants of the Canadian prairies

(Publication ; 1751)

"Replaces Publication 964, "Ninety-nine range forage plants of the Canadian prairies", by J.B. Campbell, K.F. Best and A.C. Budd." — T.p. verso

1. Forage plants — Prairie provinces. 2. Range plants — Prairie provinces. I. Canada. Agriculture Canada. Research Branch. II. Title. III. Title: 111 range and forage plants of the Canadian prairies IV. Series: Publication (Canada. Agriculture Canada). English ; 1751.

SB193.3.C3L66 633.2'009712 C83-097205-6

CONTENTS

INTRODUCTION

There are over 20 000 000 ha of rangeland covered with grass, shrubs, and bush in the prairies of Western Canada. This rangeland produces most of the summer pasturage and about 20% of the winter feed supply for about 6 000 000 head of livestock. The productivity of the rangeland is not high and varies considerably from district to district. More than 14 ha may be required to grow enough summer feed for a cow in the driest areas, whereas less than 1.5 ha suffices in the most productive areas.

The 20 000 000 ha used for range are enclosed by a semicircle that has the International Boundary between Canada and the United States as its base. Its western edge runs close to the Rocky Mountains, and the Red River in Manitoba delineates its eastern boundary. From the Foothills of the Rocky Mountains, the northern perimeter of the semicircle goes through Edmonton, Cold Lake, Meadow Lake, Nipawin, and Lake Manitoba and finishes at the Red River. In this publication this region is called the prairie area. Range comprises only about 40% of the total prairie area. The remainder is utilized for dryland farming, irrigation, and forest reserves.

More than 1500 species of native plants have been collected within the prairie area, although few are abundant and even fewer are palatable to livestock. Within any district, four or five grasses may provide 90% or more of the grazing supply, and a single grass may provide as much as 70% of the forage in small sections. Therefore, although many plant species grow in the prairie area, few are important to the livestock industry.

The plants included in this bulletin are the most economically important of the grasses, herbs not otherwise listed, legumes, poisonous plants, rushes and sedges, and trees and shrubs that occur in the prairie area. Habitat, growth characters, palatability, and nutritive value are described, and line drawings are included to aid identification.

The common names of native plants in this publication are in accordance with those recommended in *Common and Botanical Names of Weeds in Canada* by J. F. Alex, R. Cayouette, and G. A. Mulligan, prepared under the direction of the Expert Committee on Weeds, Canadian Agricultural Services Coordinating Committee (Agriculture Canada Publication 1397). This publication establishes a single, preferred common name for many commonly occurring native plants in Canada, in an attempt to encourage uniformity in useage and spelling across the country. Alternate, well-known regional names are given in the text and listed in the index. Common names for species that are not listed in the above publication follow those set out in *Budd's Flora of the Canadian Prairie Provinces,* revised and enlarged by J. Looman and K. F. Best (Agriculture Canada Publication 1662).

The useage and spelling of common names for cultivated plants sometimes differ from those for native plants, even when the same species is being referred to. The common names of cultivated plants, given in the text when applicable, generally follow the useage recommended in *Hortus Third,* revised and expanded by the staff of the Liberty Hyde Bailey Hortorium (MacMillan Publishing Co., Inc., NY), with spellings consistent with those in other Agriculture Canada publications.

Grasses

Without doubt, grasses are the most important of the forages for livestock. The seeds and juices of some grasses have also been important items in the human diet for centuries, and the stems, roots, and leaves of other grasses have provided fuel, roofing, and fiber for homes and industry since the dawn of civilization. Grasses also afford the soil some protection against wind and water erosion, one of their most important functions.

Pasture is the cheapest and most common method of utilizing grasses, but in northern districts hay may be equally important. The usefulness of grass silage has been realized for many years, and silaging has become a common method of preserving grass throughout North America and Europe. Under favorable conditions one hectare of range can produce enough forage in the form of fresh-cut grass to feed 10 or 12 mature cattle. Drying and pelleting or compressing fresh, young grass preserves feed quality and provides high protein supplements. Thus, grass is a versatile crop that can be utilized in several ways to provide food for growing and mature livestock.

There are more than 6000 species of grasses throughout the world, all of which have certain characters in common that distinguish them from other plants. Grasses have hollow or, occasionally, pith-filled stems that are divided into sections by membranes that form nodes. The leaves are composed of two parts: a sheath enclosing the stem and a free blade. The leaves grow in ranks of two, with each leaf of the pair occurring on opposite sides of the stem. Grass flowers are inconspicuous, enclosed in chaffy covers called lemmas. There may be several flowers enclosed in outer chaffs, the glumes, or there may be only a single flower. The seeds may be free of the lemmas, as in wheat, or enclosed, as in oats. Several of these distinguishing characters are presented in Fig. 1, and reference to the illustrations throughout the text may assist in identifying the various grasses.

More than 140 species of grass occur in the prairie area. Several of these species grow only under specific conditions. Blue grama occurs in the dry central region and is rare elsewhere. Rough fescue is the dominant grass in the Foothills and Cypress Hills. The bluestems are dominant in Manitoba but are rare west of the third meridian. Sand grass, sand dropseed, and Indian rice grass are found only on sandy soils. Other grasses occur only where moisture is plentiful. Therefore each district and each habitat support only a few important species that produce the bulk of the forage, and the growth habits of these predominant plants determine pasture management practices and carrying capacities in each locale.

Although most of the grasses described in this publication are native range plants, also included are a few of the common cultivated species that have been used successfully as tame pasture in range management programs on various soils and under various climatic conditions in the prairie area.

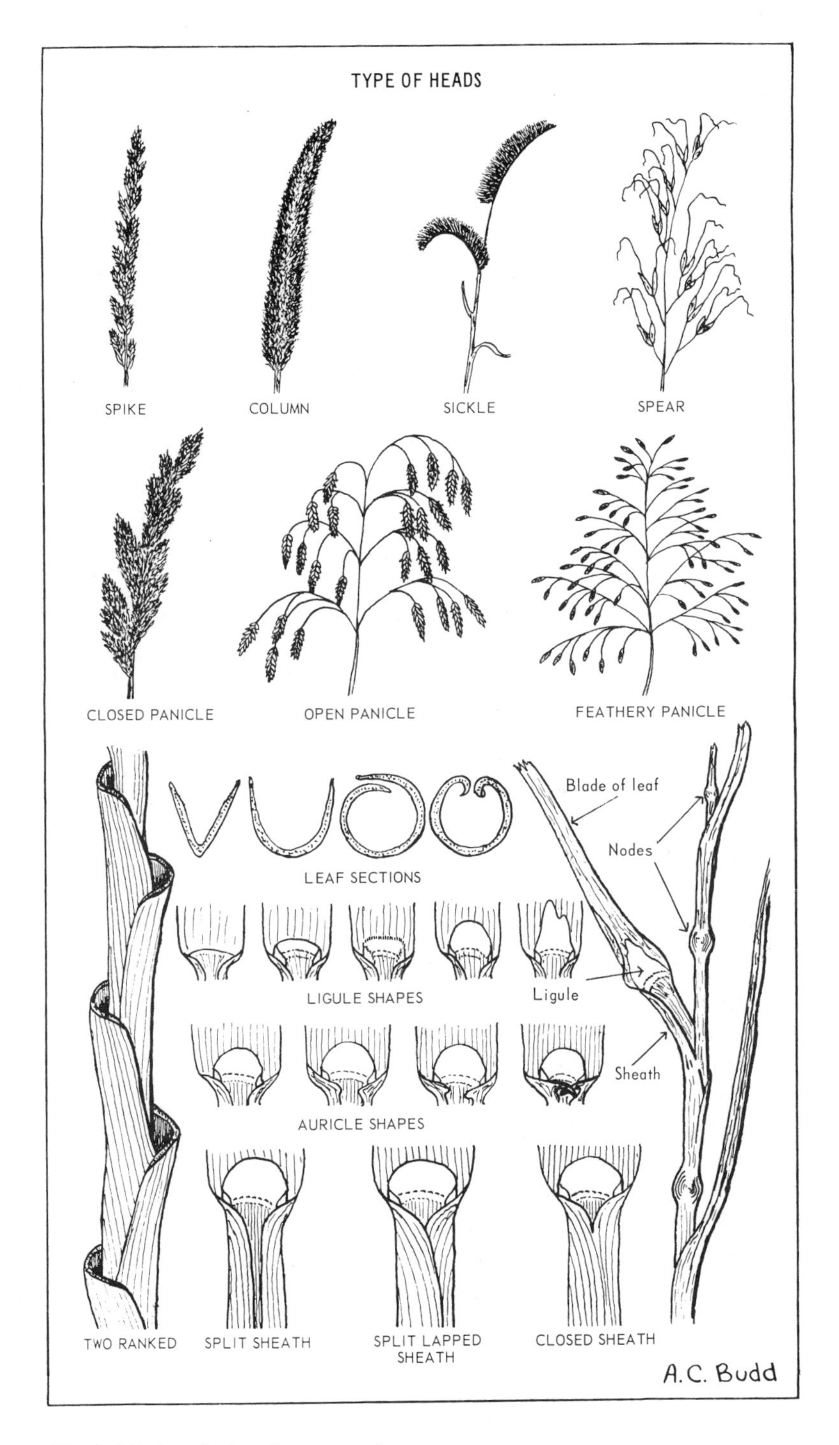

Fig. 1. Distinguishing characters of grasses.

Herbs not otherwise listed

Generally defined, a herb is any seed-producing annual, biennial, or herbaceous perennial that does not develop persistent woody tissue and that dies down at the end of a growing season. Although most of the plants in this publication are herbs by this definition, in this section we are considering certain commonly occurring range and forage herbs that do not belong to any of the other major plant groups.

The range herbs discussed in this group tend to reduce the yield of forage in pastures, whether native or seeded. Usually, however, the total amount of these miscellaneous herbs is small, and as several species are palatable, the reduction in yield is less than 10% in well-managed range. Many of the miscellaneous herbs are so small and sparse that they are not used or, when eaten, contribute little to the fodder supply. Although a few species such as the cacti can cause physical injury to grazing animals, most species are not dangerous; they are just unpalatable. Only a few are sufficiently abundant and palatable to warrant their discussion as forage plants. Such forage herbs are known as forbs.

Why certain forbs and certain browse plants are eaten readily is difficult to understand. They are classed as palatable because livestock can be seen eating them, often with apparent relish. It is easy to understand why succulent species like alfalfa and dandelion are readily grazed, but many other forbs that appear much less tasty are also grazed well. They are probably well liked because the animals can perceive that these plants are rich in minerals and protein.

The herbs described in this publication are among the most common and widespread in the prairie area. Besides providing some forage, they aid in preventing soil erosion and on occasion indicate overgrazing by increases or decreases in their abundance.

Legumes

Legumes are an important part of the vegetation for two reasons. Firstly, most leguminous species develop nodules on their root systems in which nitrogen-fixing bacteria live. Thus, legumes add nitrogen to the soil. Secondly, forage from legumes can either be palatable and desirable or potentially or actually poisonous and therefore undesirable. Few native legumes are palatable. The most important legumes for pasture and hay are introduced species.

Legumes are not only usually high in protein and calcium but also in crude fiber, especially later in the season. Early harvesting, either by allowing animals to graze or by cutting for hay, makes best use of the desirable characteristics of legumes. In pastures seeded to grasses and legumes, early grazing results in high protein content throughout the season in both types of vegetation and reduces the chances of cattle grazing on alfalfa, which causes bloat.

Poisonous plants

A considerable number of plants in native range are not only unpalatable but also poisonous or potentially poisonous. Most of these plants cause trouble only when livestock is under stress, for example after they have been herded into fresh pasture after the range has been depleted. Only a few poisonous species appear to be palatable and are eaten by livestock when not under stress. Animals often graze these species because they provide the most succulent growth among other, less palatable vegetation.

Only the species commonly found to be the cause of livestock poisoning are included in this publication. Two of these species, seaside arrow-grass and low larkspur, have accounted for 95% of cattle losses in the prairie area over many years. Sheep poisoning is usually caused by death camas, but in several areas chokecherry, and possibly also saskatoon, are found to be responsible.

Livestock poisoning is usually avoidable by good management. When grazing of pastures in which low larkspur or death camas occurs is deferred until the forage grasses are well developed, these poisonous species become less attractive to the animals. Providing water and salt near gates through which cattle are herded into pastures containing heavy stands of seaside arrow-grass can also prevent severe losses. Finally, locoweeds and milk-vetches cause trouble only when pastures in which these plants occur are overgrazed.

Rushes and sedges

Rushes and sedges resemble grasses in that many species produce long, narrow, flat, basal leaves, as well as a few stem leaves. They can be distinguished from the grasses, however, by a few simple characters: the stems are solid, without nodes; the sheaths are always closed; and the leaves occur in ranks of three. A few of these distinguishing characters are shown diagrammatically in Fig. 2.

The sedge family is one of the largest in the plant kingdom. It includes about 4000 species, some of which occur in the tropics, others in the north or south temperate zones, and a few north of the Arctic Circle or in alpine meadows. More than 160 members of the sedge family occur in the Prairie Provinces. Some of these are rare. Others grow only under very specific conditions and are abundant in only a few areas. Still others are common in upland prairie or in sloughs and at lake margins.

Certain sedges are palatable to livestock but are quickly grazed out when they are not managed properly. A few dryland sedges are so small that they cannot be grazed by cattle. One of these, which spreads under heavy grazing, is an indicator of overgrazing.

Many sedges found in the prairie area have creeping roots. The extended roots sometimes produce stems and leaves at short intervals; the plants thus appear to have a tufted growth habit. In other species the stems occur at widely spaced intervals; each stem thus appears to be a separate plant. Most species send up stems at distances of 1–10 cm and produce a fairly dense cover, resistant to erosion.

Inflorescences of sedges vary considerably. In certain species such as the thread-leaved sedge, all flowers are enclosed in a single spike at the top of a stem. Others, including beaked sedge, have two types of spikes on each

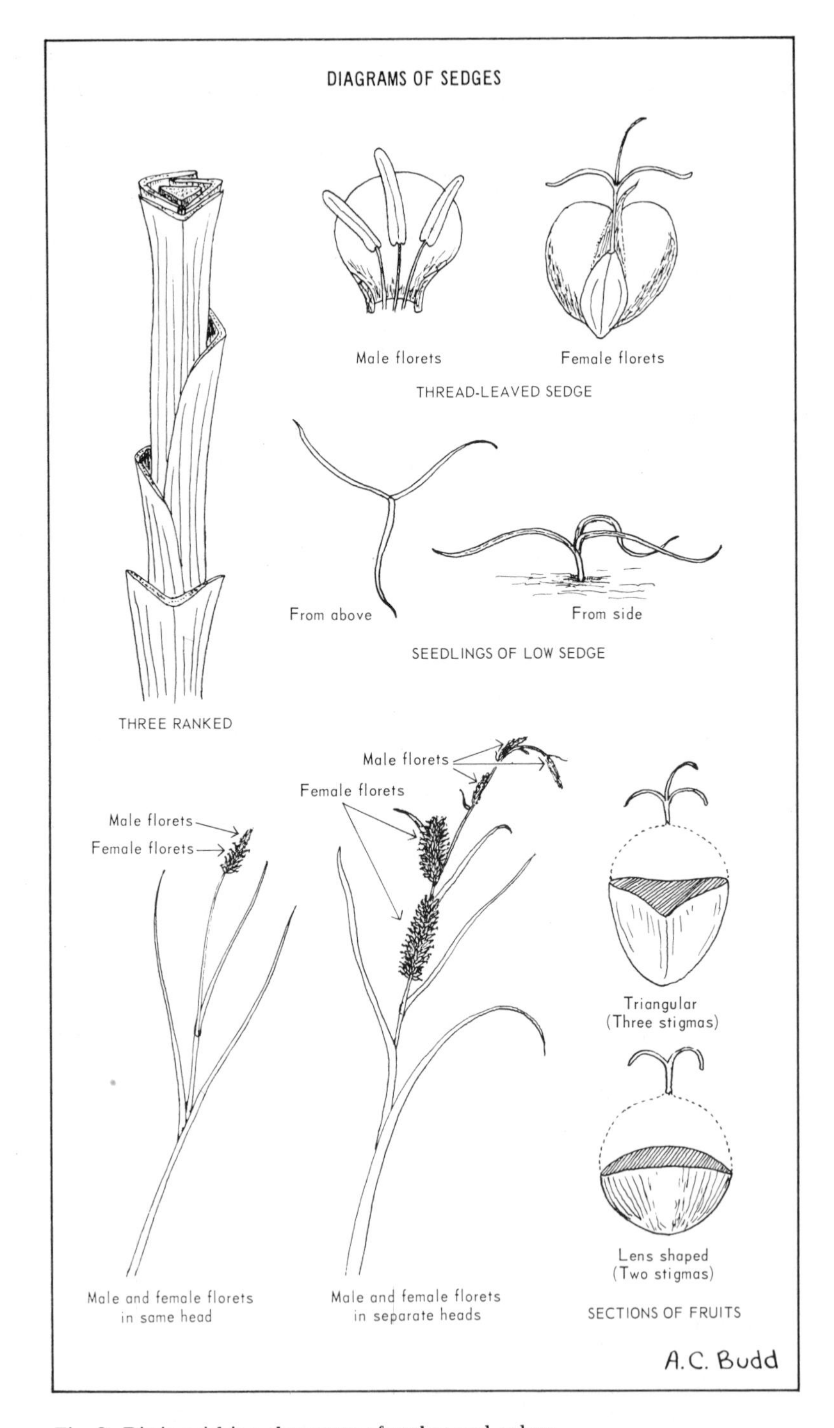

Fig. 2. Distinguishing characters of rushes and sedges.

stem: the male flowers in narrow spikes at the tip of the stem, and the female flowers in thick spikes in the axils of the leaves. A few species have male and female flowers on different plants. The seed of sedges is an achene, which may be enclosed either in a sac-like structure or between scales.

The stem of the true rushes is usually smooth and round, shiny, leafless, and filled with pith. The flowers are perfect and arranged in an inflorescence that appears to come from the side of the stem, but in fact arises from the axil of the long leaf that continues from the point where the inflorescence emerges. The small seeds are contained in a three-celled capsule.

Chemical analyses of rushes and sedges indicate compositions similar to those of grasses, with somewhat higher contents of protein, fats, digestible starches, and sugars and lower contents of crude fiber and ash. Hay of sedges and rushes is nutritious but lightweight, and palatable and unpalatable species usually grow mixed in stands.

Trees and shrubs

Most of the trees and shrubs in the prairie area have no forage value, but the young growth of a few species or green foliage late in the grazing season can provide browse. Why the browsing of trees and shrubs occurs is as hard to understand as why the grazing of some forbs occurs. One suggestion is that animals seek out certain species, such as the willows, for their medicinal value. However, why species with poisonous properties are also browsed is not clear. Trees and shrubs, like some forbs, are likely grazed because they provide certain minerals and other specific nutrients lacking in the regular fare. Chemical analyses reveal higher protein and phosphorus contents in the new growth of shrubs than in the associated grasses. This suggestion is further substantiated by the increased utilization of forbs and browse when nutrient contents of grasses decline.

However, browse and forbs cannot replace grasses in livestock rations. In sparse stands sagebrush, roses, willows, poplar, and other species may be grazed out, but when the stands of these trees and shrubs are even moderately heavy, these species may increase in abundance as grazing reduces the stands of grass.

Grasses

Creeping bent grass *Agrostis stolonifera* L.

Creeping bent grass, also widely known in the prairies as redtop, was imported from Europe in the days of early settlement and probably prior to 1750 A.D. It spread rapidly throughout North America and now is well-established wherever abundant moisture supplies are available. In the Canadian prairies this species often occupies slough edges between the upland grasses and the sedges growing in the water.

Creeping bent grass forms a dense turf. Short creeping roots spread its stand, and dense, shallow feeding roots provide abundant nutrients. Its leaves are short and mostly basal. The spreading panicle bears many tiny, reddish flowers from which the name redtop is derived. Small, numerous hard seeds are matured early in summer.

The palatability of creeping bent grass has been rated low, but whereas most grasses become less palatable late in the season, creeping bent grass is better utilized from July on. Its protein content drops from about 22% in May to 16% at maturity, and it has a good nutrient balance.

It grows best on poor, rather wet soils, and once established can produce a good crop of hay. It also is an excellent soil binder wherever it grows.

K. F. Best

Rough hair grass *Agrostis scabra* Willd.

Rough hair grass, which is related to creeping bent grass, is often called winter redtop as well as tickle grass. It has a widespread distribution throughout Canada and the United States. It prefers to grow on moist, well-drained soils. In the prairie area only waste places, dry sloughs, roadsides, and abandoned land support its existence. It is one of the first grasses to show green in the spring.

Rough hair grass is a short-lived perennial with a fibrous, shallow root system and short leaves, both growing from a crown. Its seed stalks are 20–50 cm tall and are each topped with a large, finely branched panicle. When walking through a stand, one often finds the panicles breaking off and creeping up inside one's trousers. The panicles are also broken off in strong winds and can be seen tumbling across the fields.

Plants of rough hair grass emerge in mid-April and make a leafy growth. Heads emerge from the sheath in late June, and seed matures from late July to mid-August. By September the plants dry up, but they can still be recognized by the small, straw-colored clumps dotting abandoned fields, dry sloughs, and waste places.

Only during April and May is this plant palatable to livestock. Animals refuse to eat it after the shot-blade stage, although the basal leaves may remain green and nutritive until late July. Its only other value is its ability to colonize denuded soils, thus assisting to prevent erosion.

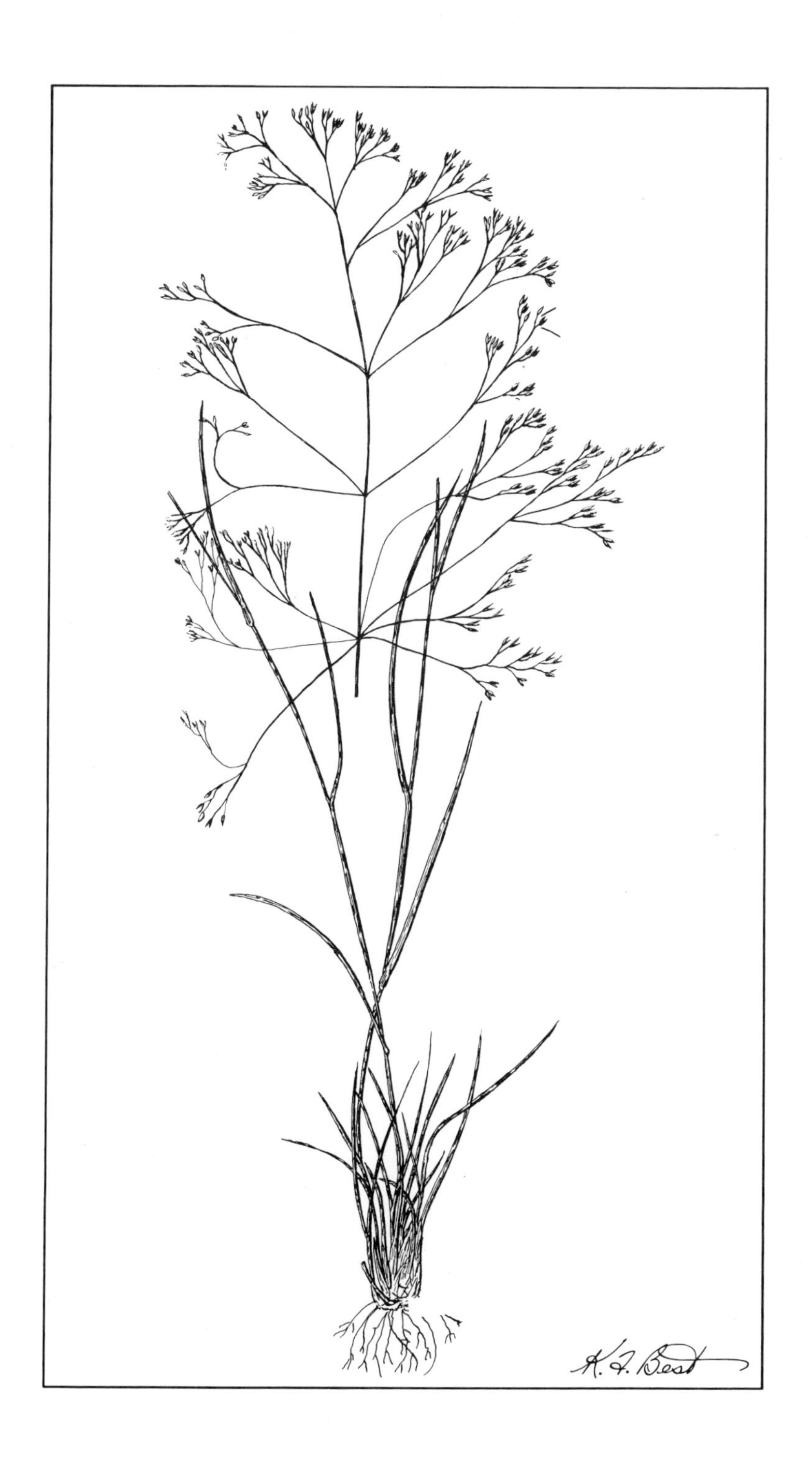
K. F. Best

Blue grama *Bouteloua gracilis* (H.B.K.) Lag.

Grama grasses are southern species. The only two that extend into Canada are blue grama, which is very common throughout the Prairie Provinces; and side-oat grama, *B. curtipendula* (Michx.) Torr., which occurs sparsely in the Souris River basin. The name grama is a literal translation of the Spanish for grass.

Blue grama is the most abundant grass in the short-grass prairie. Its range extends across North America and southward through Mexico into South America. It becomes progressively sparser northward but is quite common as far north as the tree line across Alberta and Saskatchewan. It starts growth late in the season, about 5 May in southwestern Saskatchewan and nearly a fortnight later at its northern limit of growth.

Blue grama produces a dense mat of short leaves that grow from short, underground tillers as well as from a crown. The pith-filled stalks grow up to 50 cm tall and each usually carries two dark brown, sickle-shaped spikes with all the flowers clustered along the upper sides. It requires about 75 days to mature seed. It produces a mass of surface roots that extend down about 50 cm; a few secondary roots feed down to about 1 m.

Under the best growth conditions of short-grass prairie, an average stand of blue grama yields only 100–125 kg/ha a year, which comprises about 35–45% of the total forage production. During midsummer it does not appear to be particularly palatable, but it is readily eaten during the autumn and early winter. It is high in digestibility and it cures on the stem. It recovers rapidly from drought and is quite resistant to grazing and trampling. Chemical analyses indicate a well-balanced nutrient content, with protein high during the spring but dropping to about 5% when the grass is cured.

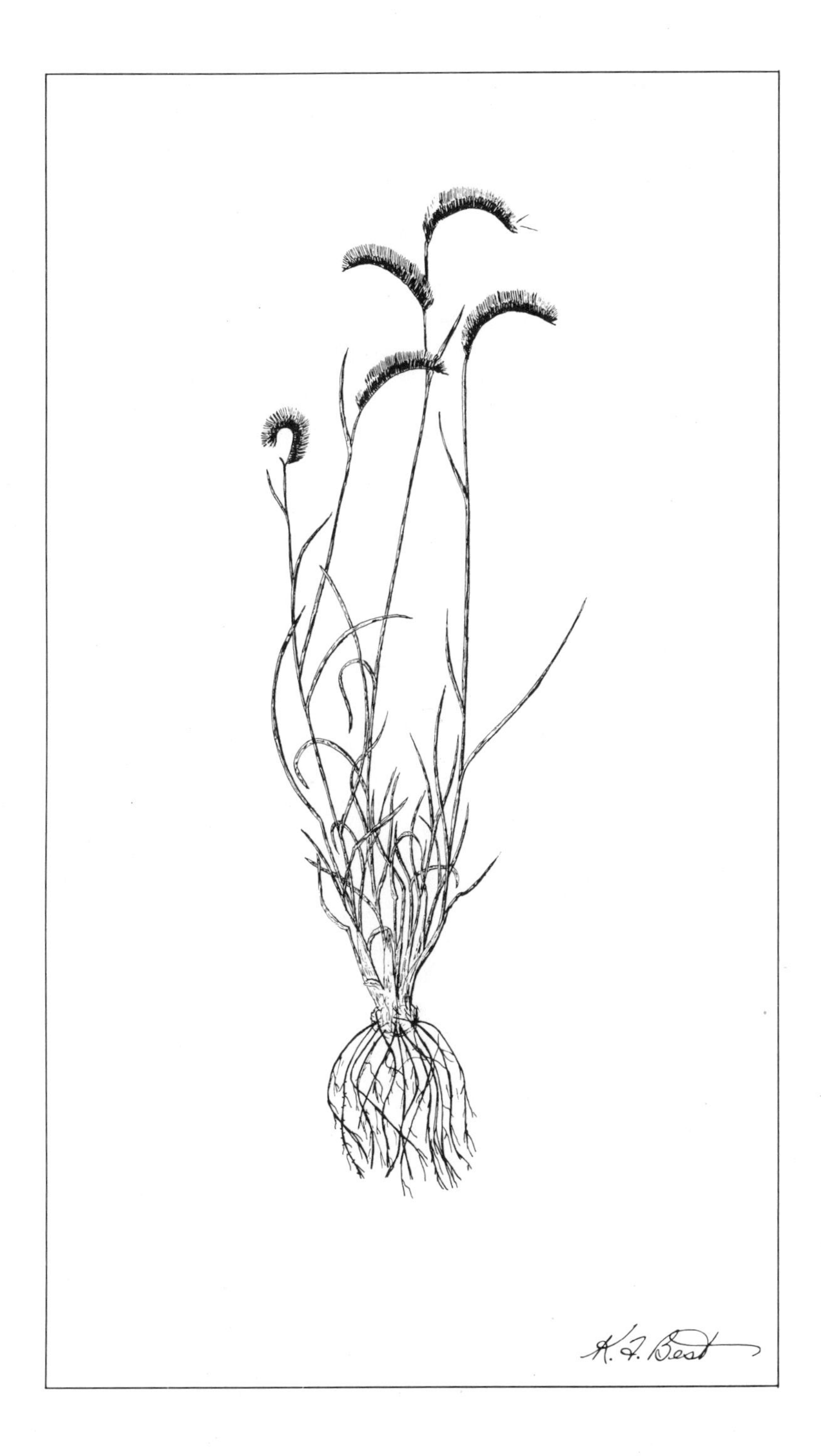
K. F. Best

Canada blue grass *Poa compressa* L.

The blue grass group, *Poa*, is one of the largest among grasses. Species are distributed throughout the world from sea level to high elevations in the mountains, and in dry, moist, or saline sites. The group is distinguished from other grasses by its boat-shaped leaf tips. Of the nearly 70 species native to North America, some 16 grow in the prairie area; of these, five are important forage grasses, although none are high yielders.

Canada blue grass is a native perennial known also as flatstem blue grass and Virginia blue grass. It has wide distribution throughout all of Canada and the United States, and it is an important cultivated grass in Eastern Canada and the states of New York, Pennsylvania, and Maine. In the prairie area it is found throughout western Manitoba and locally in central Saskatchewan and Alberta. It grows on a variety of soils of low fertility or with poor drainage characters. Its ability to persist and produce on poor-quality soils has made it popular for regrassing eroded areas. It requires more moisture than is available in the dry central portion of the prairie area.

Canada blue grass has a dense creeping root system, as well as numerous fibrous roots that extend deep into the soil. Its numerous basal leaves are short, folded when young, nearly flat at maturity, and boat shaped. The short stems are flat or compressed and end in a narrow panicle. Numerous seeds are produced, each with a few cobweblike hairs at the base.

The short, dense basal leaves of Canada blue grass are palatable from early spring until late fall. Under some conditions of growth Canada blue grass makes fairly productive hay, but it is better known for its good resistance to grazing and tramping. In certain areas it is used as a lawn grass and for golf fairways. Occasionally, it is found growing in mixtures with Kentucky blue grass.

Spring protein content of Canada blue grass is high, 18–22%, but the natural seasonal decline reduces this level to 4–5% by October. Canada blue grass shows some natural curing properties, in that livestock, particularly horses, can maintain a good condition when pasturing on it during the autumn and early winter.

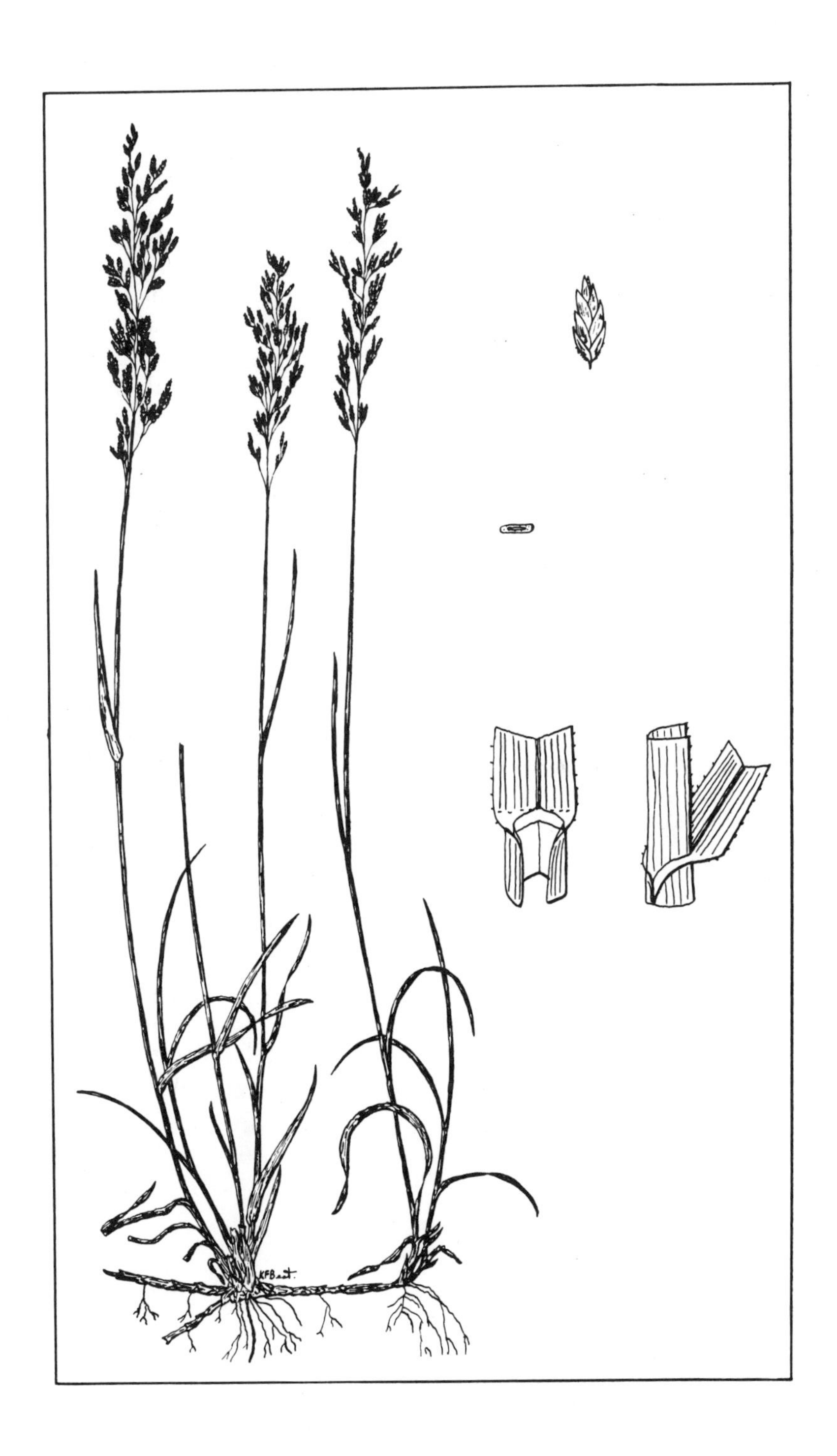
KFB del.

Canby blue grass *Poa canbyi* (Scribn.) Piper

Canby blue grass is a native bunch grass. Its range includes most of Western Canada and the states west of the Mississippi River. In the prairie area it is common on slightly saline meadows in the dry central part of the region, but it can grow sparsely in dry sloughs and in slightly alkaline marshes elsewhere.

Canby blue grass has a relatively shallow root system, not over 60 cm deep. It produces considerable light green, flat, soft basal leafage, often covered with gray patches of fungus. The straw-colored stems are seldom over 60 cm tall, each ending in a short, dense, closed but branched panicle. The seeds are narrow and about 6 mm long, with a shiny straw-colored hull.

Canby blue grass is fairly palatable during early growth, but because it grows in fairly open stands it does not produce much forage. It invades irrigated alfalfa or alfalfa-grass stands where it adds very little bulk and detracts from the nutritive value of the feed. Its protein content drops rapidly from a high of 20% in early spring to 10% by the end of June, and to lower levels late in the season. Its phosphorus content is above average throughout the growing season.

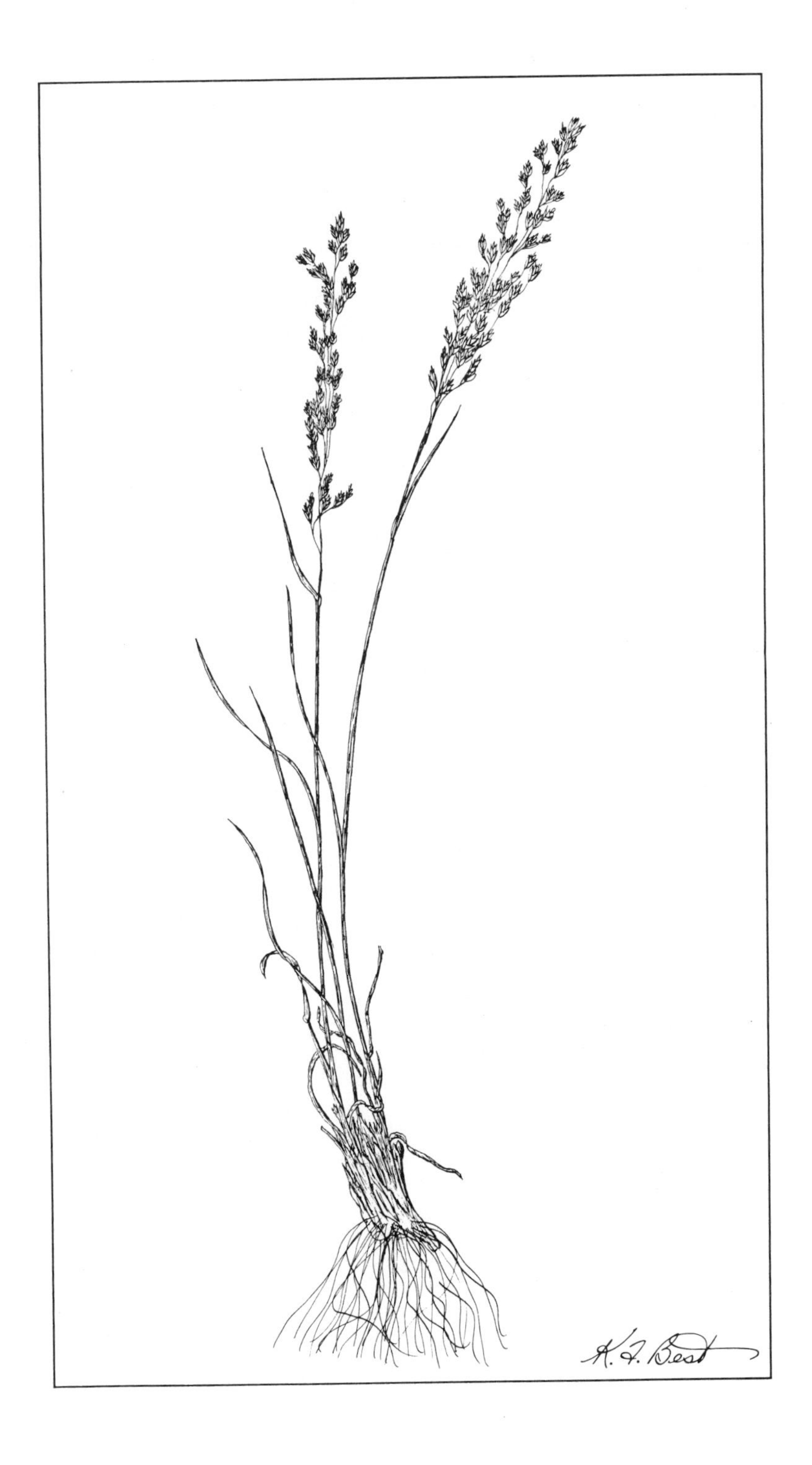
K. F. Best

Early blue grass *Poa cusickii* Vasey

Early blue grass is a native of western North America, extending westward to the Pacific Coast from Saskatchewan and southward from the Yukon to California and Colorado. Although rarely found in the extremely dry central portion, it grows throughout the prairie area. It seldom grows in heavy stands, but rather as individual plants or in small groups among the grass and shrub cover.

Early blue grass is a small bunch grass with short, dense roots that seldom penetrate more than 30 cm into the ground. Numerous short, very slender leaves grow from the crown, as well as several nearly leafless, 40-cm stems. The seed head is a straw-colored, dense, closed panicle, which produces numerous shiny seeds about 6 mm long. During the short flowering period in early June, the head enlarges considerably and becomes almost silvery in color; the large, silvery heads make the plant easy to recognize.

Early blue grass is palatable only during the spring and early summer, after which it dries up and often withers away. However, during the early season, its rapid growth provides some forage and its soft, lush leaves are attractive to livestock. Protein content is as high as 25% in spring but decreases rapidly to 10% by mid-June and to as low as 4% by August.

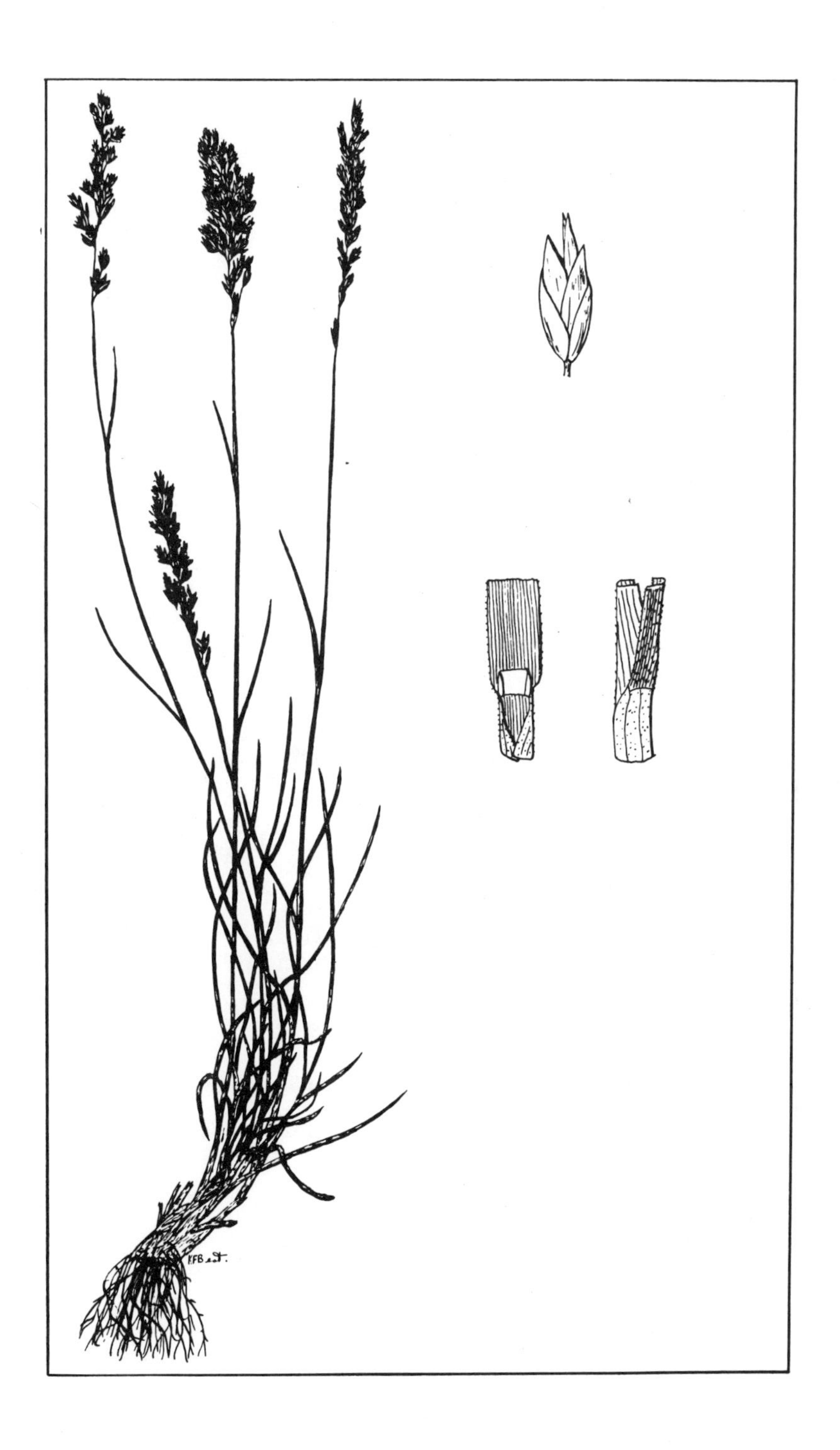

Kentucky blue grass *Poa pratensis* L.

Kentucky blue grass was introduced into America from Europe before 1700 but made slow progress in becoming established for nearly 50 years. However, as settlement moved westward from the New England states, this grass followed so rapidly that the Indians referred to it as the "white man's foot grass." Today the species is widely distributed throughout North America and has invaded rangelands to such a degree that it is often considered to be a native. In the prairie area it is an important species, particularly in western Manitoba and throughout the transition zone between prairie and forest in central Saskatchewan and Alberta. Elsewhere it occurs in sparse stands along coulees, in brush, and in other protected places.

Kentucky blue grass has dense creeping roots that extend its stand, as well as deep feeding roots. Its short, boat-shaped leaves are mostly basal, although a few occur on the 1-m-high stem. The flower head is an open, pyramidal panicle, usually with five branches at each node; of these, the center and outer branches are long and the others short. Seeds are numerous and each has a mat of cobwebby hairs at its base.

Kentucky blue grass is considered one of the best pasture grasses in eastern North America. However, it requires more moisture than is available in the prairie area and therefore has never been popular for pasture where rainfall is less than 500 mm annually. It is the grass commonly used for lawns in Western Canada. Cultivated, it is sold as Kentucky bluegrass.

During the spring Kentucky blue grass is palatable and very nutritious. However, unless it is kept grazed down, it soon becomes stemmy and unpalatable. Unless the ground is fairly moist all the time, the stands become patchy. Protein content may rise to 30% briefly during the spring but it drops rapidly as heading occurs. When this grass is grazed continuously, the regrowth maintains a high protein content.

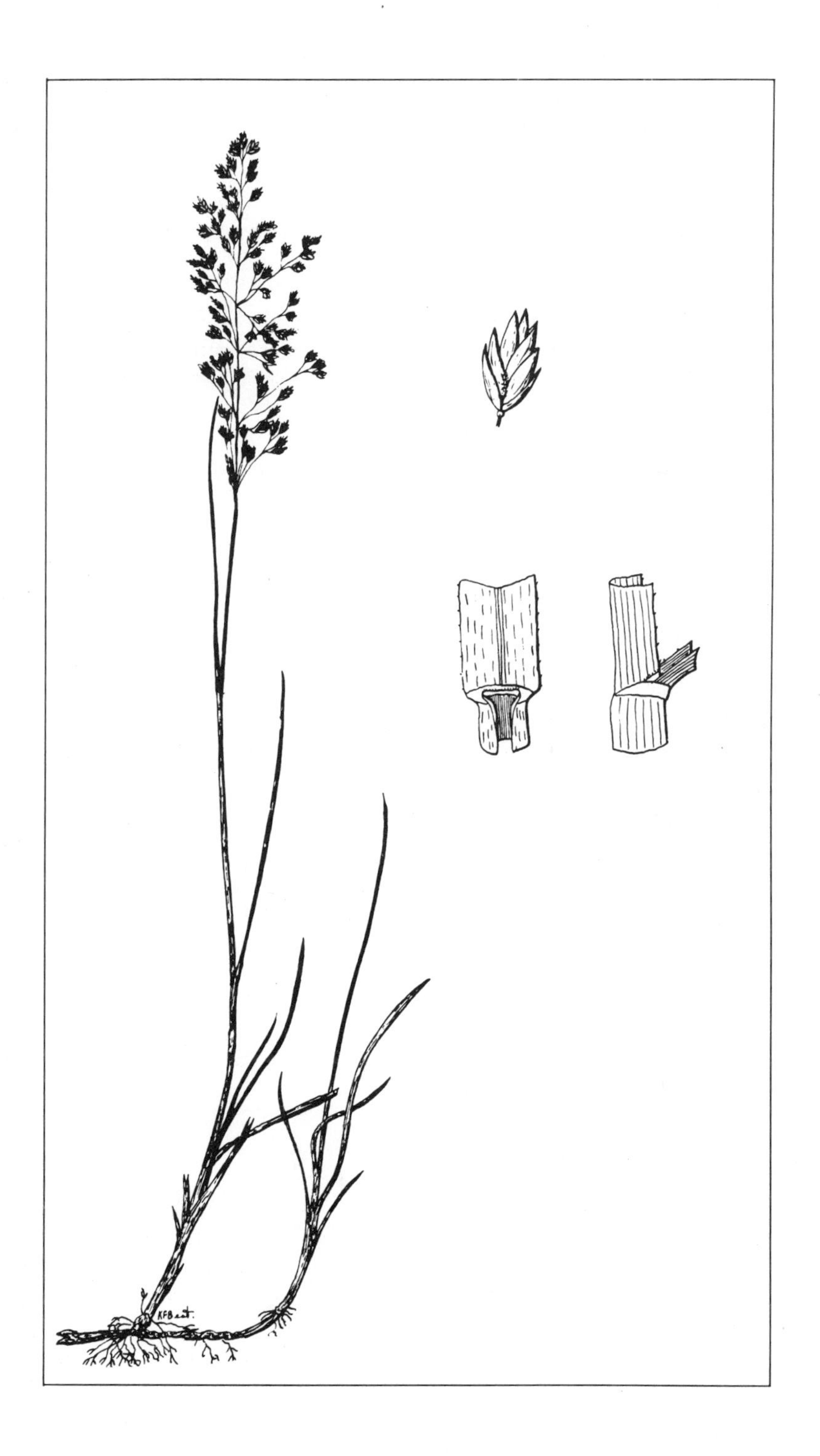

Sandberg's blue grass *Poa sandbergii* Vasey

Sandberg's blue grass is a common native that grows throughout the northern Interior Plains and westward across the Rocky Mountains to the Pacific Coast. Its southward extension includes Mexico. It is found throughout the prairie area, but it is rarely very abundant. Locally it may form 7–8% of the total cover, especially after severe overgrazing or disturbance of the range.

Sandberg's blue grass is a small bunch grass. Its roots seldom penetrate more than 40 cm into the ground. It has numerous fine, folded basal leaves and a short stem that may reach a height of 30 cm. Its seed heads are closed panicles, which are straw colored and shiny when mature.

Sandberg's blue grass grows rapidly in late April and early May, growth often beginning even before the needle grasses and western wheat grass show green. It flowers in late May or early June and matures its seed by early July. Its leaves then wither into whitish clumps that remain dormant until September, when some new growth may occur. It is drought resistant and during dry years may spread as less drought-tolerant species die back.

By no standard can it be considered a good forage grass, because its leaves are not very palatable and it is a low forage producer. By mid-June all stock avoids it, and after this date animals invariably drop this plant from their mouths when they inadvertently graze it. Some authorities consider the plant to be a pasture weed because it can increase its stand as better grasses are killed out. During April its protein content may be as high as 25% but it is reduced to 5% by early July. Its extremely early growth, producing greenness when all other grasses are still dormant, help to identify Sandberg's blue grass.

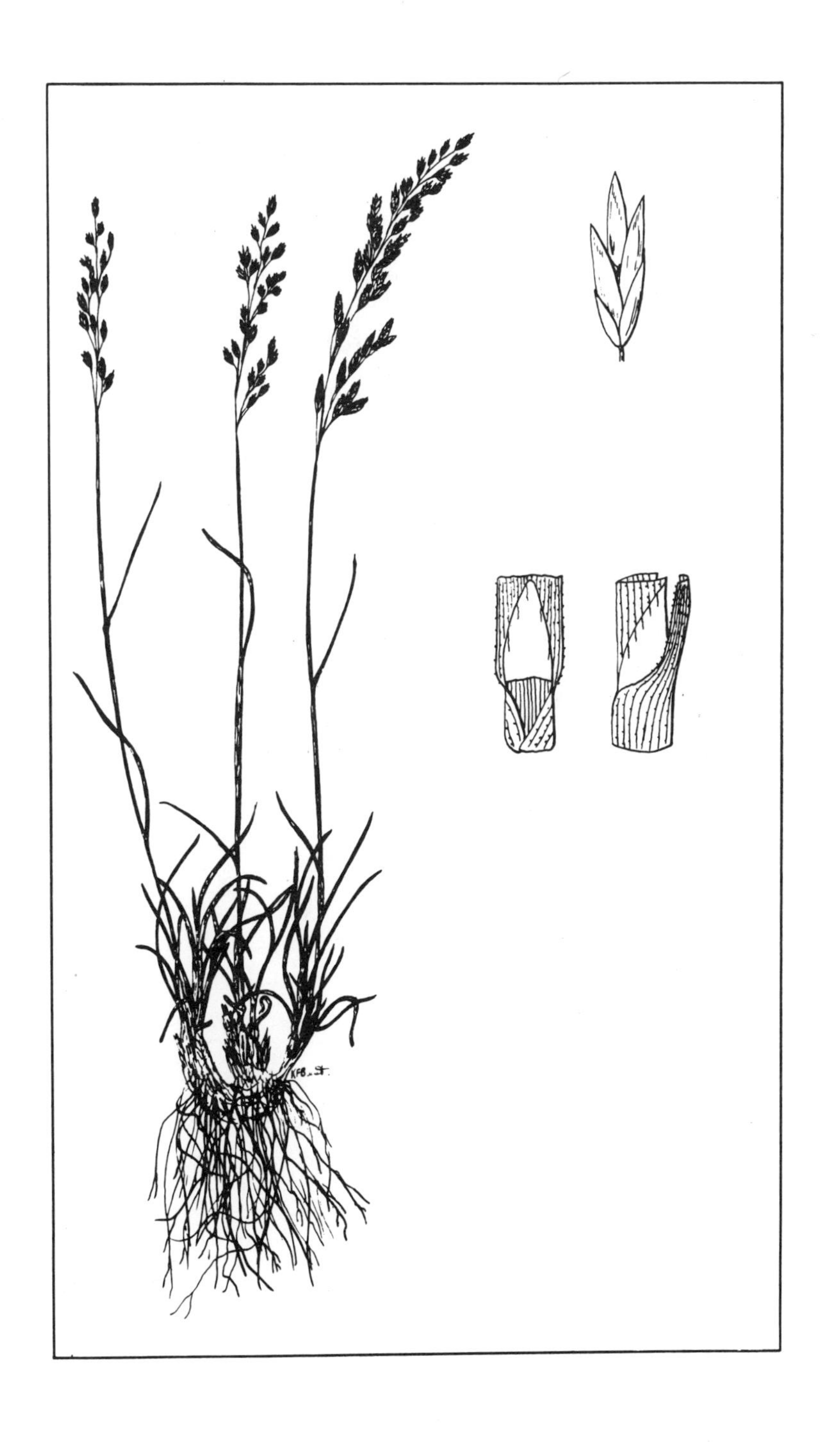
KFB-St.

Big bluestem *Andropogon gerardi* Vitm.

The North American home of the bluestems is the region surrounding the Gulf of Mexico. A few species extend their ranges up the Mississippi River system and the east and west coasts, but only two appear north of the International Boundary. The generic name *Andropogon* is derived from two Greek words meaning man's beard and refers to the long, white hairs contained in the head.

Big bluestem, also called bluejoint turkeyfoot or northern beard grass, grows in the eastern part of the prairie area. Heavy stands occurred in the days of first settlement throughout the Portage Plains and the Red River Valley. Today relic stands occur in the same districts, as well as sparse stands along the Souris and Assiniboine rivers, and less frequently in the surrounding uplands.

Big bluestem produces from one to many bluish stems growing to a height of 1-2 m. These robust stems are filled with pith, like cornstalks. Its heads each have two to six branches, all arising from one point and somewhat resembling a turkey's foot. Numerous white hairs fill the spaces between the seeds, and twisted awns extend from the seed husks.

Big bluestem is nutritious and palatable. During days of early settlement in the Red River Valley it produced a large portion of the pasture and hay supply. However, it was susceptible to grazing and its heavy stands disappeared as the grazing load increased; its disappearance accelerated as the land was brought under cultivation.

K. F. Best

Little bluestem *Andropogon scoparius* Michx.

Little bluestem has many common names, including prairie and brown beard grass, broomsedge, and small feather grass. Its range extends from the Rocky Mountains to the Atlantic Coast and from the Gulf of Mexico to the southern Yukon. Dense stands occur from the interlake area in Manitoba to the Manitoba–Saskatchewan boundary, and small, sparse stands may be found throughout the balance of the prairie area on sandy and gravelly soils. It is fairly common in the Great Sand Hills, throughout the Missouri Coteau, and along the South Saskatchewan and Frenchman rivers and Swift Current Creek.

Little bluestem is a bunch grass producing many pith-filled stems that can reach a height of 60 cm. Numerous green or bluish, hairy leaves grow from the crown to form a heavy mat in thick stands. A single, branched panicle tops each stem and produces the hair-covered and awned seeds typical of the bluestems. After frost the leaves develop a reddish tinge.

Although chemical analyses indicate satisfactory nutritive qualities, little bluestem is unpalatable except during the spring, when its leaves are succulent and tender. After the boot stage this plant is seldom grazed, and as it does not cure but rather withers away in the fall, it has little to recommend itself for fall and winter pasture. Its protein content drops rapidly from a high of 14–16% during mid-May, to 4% by mid-August, and to still lower levels by October. At no time during its growth does it have a protein content equal to that of the high-quality range grasses. In certain sandhill regions it is considered to be not only worthless but a pest. Experience in the interlake region of Manitoba indicates that it may be a valuable spring fodder, provided the old, dry growth is burned off once every 3rd or 4th year.

Fringed brome grass *Bromus ciliatus* L.

Brome grasses are temperate zone grasses. Their name is derived from the Greek word meaning food. They have a common character in that the edges of the leaf sheath are grown together to form a tube. Nearly 40 species are natives of North America.

Fringed brome grass is a native whose range extends throughout southern Canada and northern United States. It possesses characters similar to several native brome grasses which are difficult to distinguish from each other and which have closely related nutritive values. Fringed brome grass and its associates grow in moist sites at the perimeter of the prairie area and at elevated sites within the region.

Fringed brome grass is a perennial bunch grass, as are most of the native species. Its deep and extensive, but not dense, root system supports 1-m-tall leafy stems. The panicle is open and composed of many seeds whose husks are covered partly or completely with short hairs. Fringed brome grass itself has no awns on its seeds, but some of the closely related species possess this character. Growth usually commences about mid-May.

Fringed brome grass is palatable to all classes of livestock from date of emergence until maturity. Sheep eat the ripe seeds with relish, and cattle and game strip the dry leaves from the coarse stalk. Brome grasses as a group do not cure on the stem, but all produce excellent hay when cut while green and palatable. Chemical analyses show fringed brome grass to have a May and early June protein content of 20% or more, which drops to 9–10% by early July and to as low as 3% by October. Digestible carbohydrate content increases rapidly until early August but declines as the plant dries and the leaves fall off.

A related species, downy brome, *Bromus tectorum* L., is an annual. It was introduced from Europe and has spread over range, pasture, and forest through the western United States and southern British Columbia. Downy brome is common only in the southwest corner of Alberta, but isolated small stands or plants can be found in most of the southern prairie area.

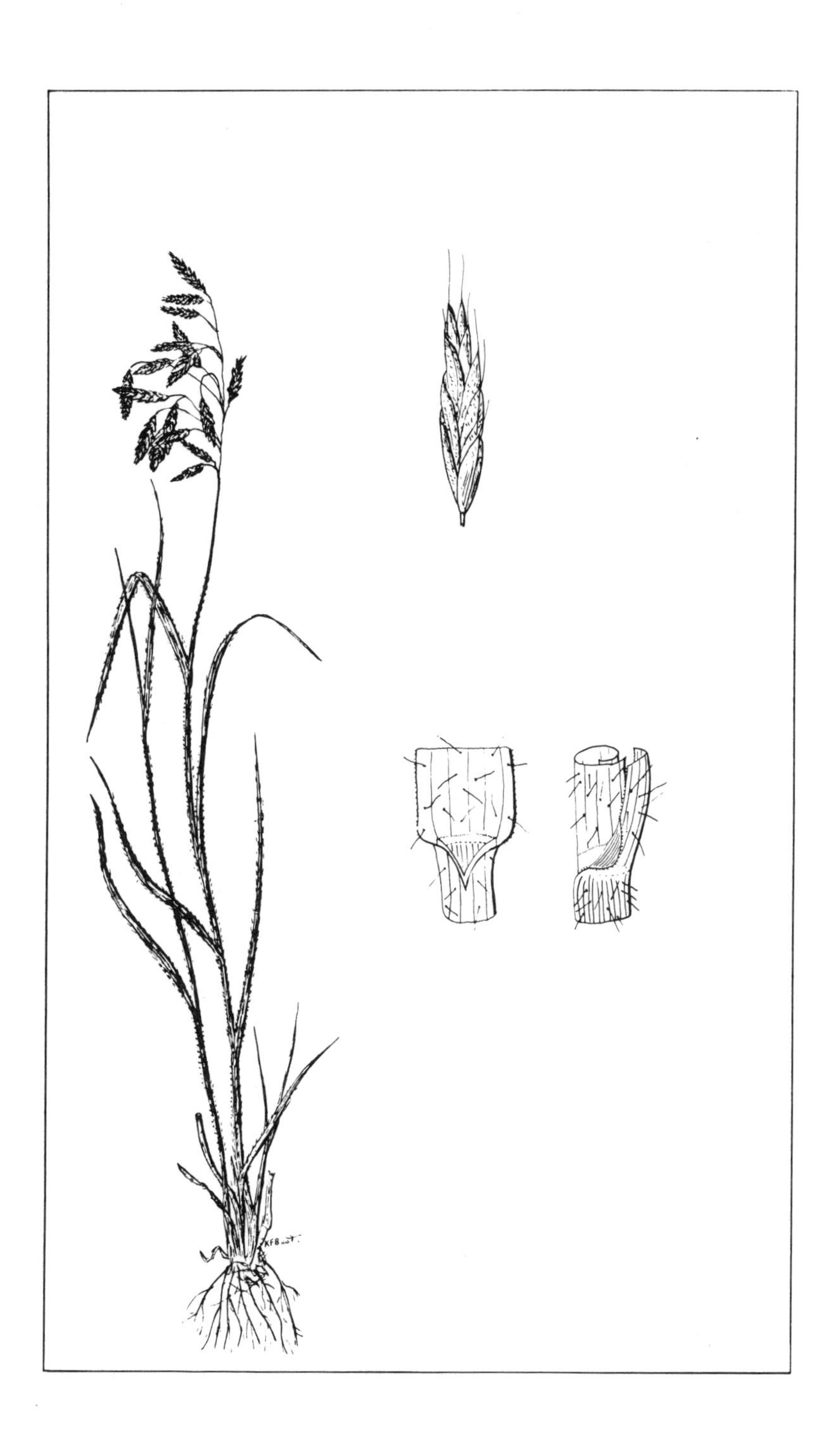

Meadow brome grass

Bromus biebersteinii Roem. & Schult.

Meadow brome grass is a native of southeastern Europe and the adjacent Near East. It is an important grass of the moist steppes and montane grasslands up to 1500 m altitude, on Dark Brown and Black soils in areas where precipitation approaches 500 mm. It resembles smooth brome but has only short or no rhizomes. Also, its native habitat is somewhat warmer and drier than that of smooth brome.

Meadow brome grass is densely tufted. The stems may reach 60–90 cm in height; the leaves are commonly up to 20 cm long and about 3 mm wide. Sheaths and leaves are usually pubescent. The seed head is 10–20 cm long, and the seed 10–12 mm long.

Varieties obtained from this species are available for cultivation and are useful for irrigated hayfields as well as for pasture in areas with more than 380 mm of precipitation.

R.M.B.

Smooth brome *Bromus inermis* Leyss.

Smooth brome is a perennial pasture and hay grass imported from Europe. Its home is in the meadows, ungrazed steppes, parklands, and open woods of eastern Europe and Siberia, and it is well adapted to similar habitats in the northcentral United States and Western Canada.

Today, sold as awnless bromegrass, it is one of the best cultivated grasses for the Dark Brown and Black soil zones of the prairie area.

Smooth brome has creeping underground stems or rootstocks from which the 1-m-tall leafy stems arise. Its root system penetrates to a depth of 1-1.5 m but only in the top 30 cm does a dense mass of roots develop. Both basal and stem leaves are broad, long, and tender; the sheath forms a tube. The panicle-type head produces a heavy seed crop, 650-750 kg/ha being not uncommon. Individual seeds are enclosed in brownish husks that may have a short awn up to 3 mm long.

Smooth brome is very palatable to all classes of livestock from emergence to the heading stage, after which only the basal and stem leaves and the seeds are eaten. A process known as lignification progresses rapidly as the stem develops, making the stem harsh, brittle, and undigestible. Growth commences early in May, and in early stages of growth this forage may have a protein content of 20-22%, which drops to 4-6% by the end of September. As the protein content drops, the fiber increases from 18 to 35%, while available carbohydrates increase slightly from about 45 to 50%. Digestibility drops from about 60% in early stages of growth to 40% or less at maturity.

Alkali cord grass *Spartina gracilis* Trin.

Cord grasses are found throughout the world. They have taken their name *Spartina* from the Greek word spartine, which was a type of tough cord. Because certain species are very salt tolerant, cord grasses have been planted in the estuaries of rivers where they help to bind the sand and actually build land by catching floating soil and debris.

Alkali cord grass is a native whose range extends between the Red and Mississippi rivers and the Pacific Coast. In the prairie area it is found in sandy, saline flats, around the borders of alkali sloughs, and along shallow streams where drainage is poor. The largest-known stands grow on the extensive flats of the Great Sand Hills in western Saskatchewan.

Alkali cord grass has strong, scaly creeping roots, as well as a deep feeding root system. Basal leaves are long and tough, with tiny, saw-like teeth along their edges. The seed stem is up to 1 m in height. The flower head consists of four to eight spikes that cling to the wavy stem. Each spike consists of up to 40 pairs of closely packed spikelets, all growing on one side of the spike and away from the stem.

Alkali cord grass is not particularly palatable in any stage of growth. Its protein content is low at all seasons, only about 6%, but its fiber content is very high, 40–41% at maturity. Despite its harsh characters it is used as hay in areas where it abounds, and livestock winter on it fairly well when it is supplemented with grain and alfalfa.

Various attempts have been made to regrass alkali cord grass flats with more palatable forages. However, no other species has established equivalent stands, and in fact very few of the cultivated grasses are able to grow under the saline conditions where alkali cord grass thrives.

Prairie cord grass *Spartina pectinata* Link

Prairie cord grass is a grass of fresh water marshes, although there are stands extending into brackish marshes along the Atlantic Coast. This grass has a limited range within the prairie area. It develops almost pure stands in marshes along the Red River and its tributaries and occurs in mixed stands with big bluestem, switch grass, and Indian grass on adjacent uplands. It is common in the interlake region in Manitoba. A few small stands occur in Saskatchewan as far west as Swift Current.

Prairie cord grass develops thick stands from thick, scaly creeping roots. Its feeding roots are dense and may extend 2 m into the ground. Dense stands of basal leafage are often 60 cm tall. The leaves themselves are wide at their base but converge to a pointed tip; their edges are rough, and as the leaves dry up the edges roll inward to form a tube. The heads consist of 10–20 spikes that cling to the main stem. Each spike is composed of up to 40 pairs of spikelets, all growing on one side of the spike and away from the principal stem. Awns extend from the seed hulls to give a somewhat bristly appearance to the spikes.

Prairie cord grass is seldom grazed when other forage is available. However, at least the upper parts of the leaves are grazed while stock seek food in the shallow water where it grows. Large quantities are cut for hay in central Manitoba, but this hay requires grain supplements to keep livestock healthy.

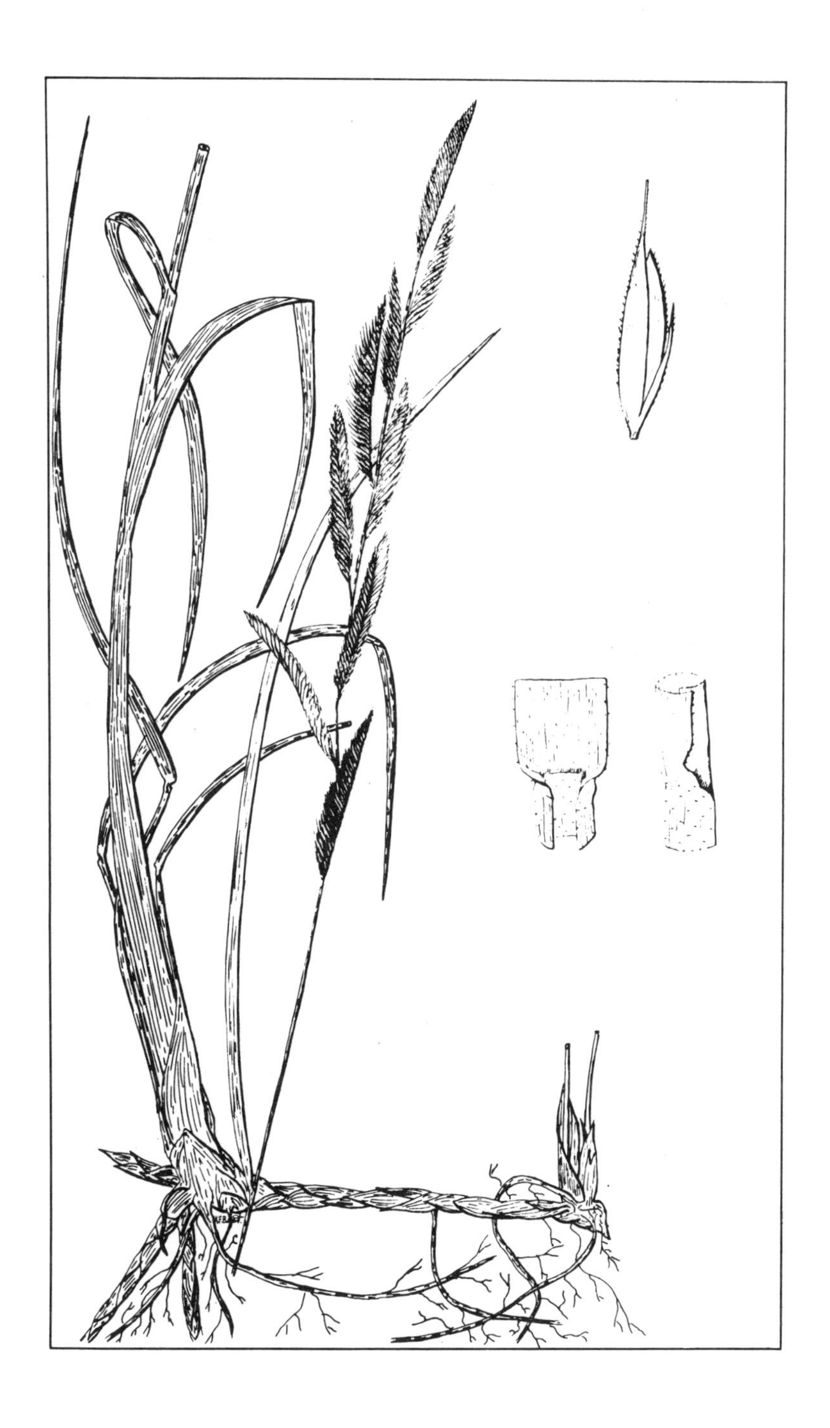

Desert salt grass *Distichlis stricta* (Torr.) Rydb.

Desert salt grass, or alkali grass, is the only species of the salt grass genus that is found in the prairie area. Its range extends from the Pacific Coast to the Red and Mississippi rivers, and southward from the northern boreal forest through the United States and Mexico. Its habitat includes all moderate saline or alkaline soils, particularly around sloughs, throughout flats, and occasionally on upland. Soil analyses at sites where desert salt grass is abundant show that the surface layers contain 0.3-0.6% soluble salts, whereas soil samples from 60-cm depth may contain more than 2%. These concentrations are sufficient to reduce growth of all cultivated crops and most natives, even under irrigation.

Desert salt grass grows from scaly, yellowish creeping rootstocks. It has a relatively dense, shallow root system, with a few deep feeding roots. There are few basal leaves, but numerous short, leafy stems occur, each of which may or may not end in a flower head. Male and female flower heads are on different plants, but both usually grow in the same clump; the male flower clusters are more numerous, larger, and a deeper yellow than the female.

Desert salt grass is not eaten when other forage is available because the plant is tough and wiry during all seasons of growth. When other forage is not available, however, livestock graze desert salt grass readily and can make fair to good gains on it. The tough root system is resistant to tramping, and stands are maintained even when heavily grazed. Desert salt grass is an important grass for the prevention and control of erosion on saline soils. Chemical analyses show a well-balanced nutrient composition. Protein and phosphorus are average, crude fiber is low, and available carbohydrates are relatively high.

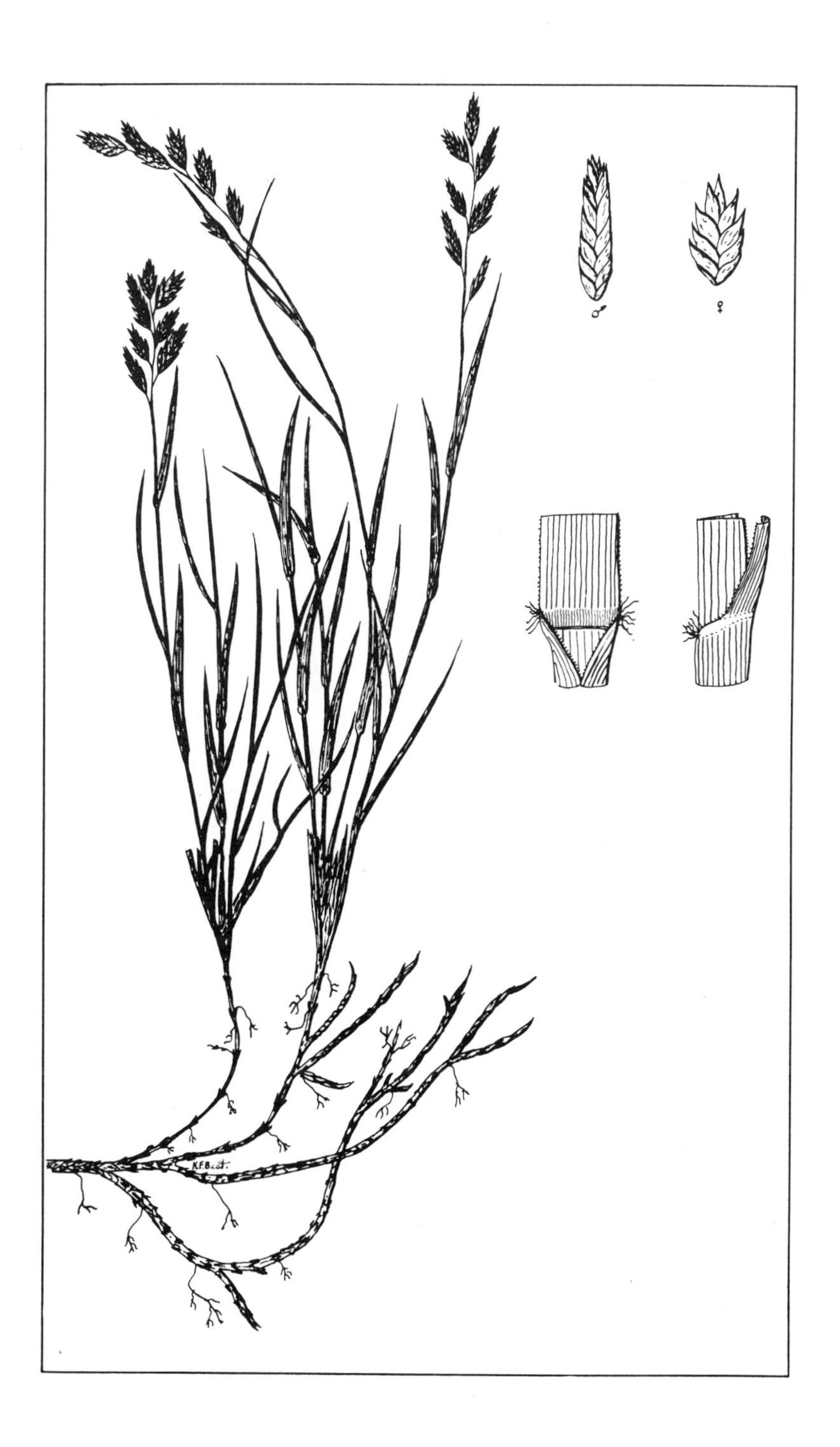

♂
♀
K.F.Best.

Bluebunch fescue *Festuca idahoensis* Elmer

There are over 100 species of fescue growing in temperate regions throughout the world. A few of these have been used in cultivated pastures throughout Europe for many centuries. Three native and two introduced species are important grasses in the prairie area of Canada.

Bluebunch or Idaho fescue is a native species whose center of distribution is western Montana. It extends northward through the Foothills in Alberta to north of Calgary, and eastward along the Milk River Ridge and into the Cypress Hills. It grows in extensive colonies in open pine and poplar forests, on exposed benchlands, in meadows, and under many other conditions at elevations of between 900 and 2100 m.

Bluebunch fescue is deep rooted. Its dense, basal leafage is fine, long, and blue colored. It spreads by tufts growing at the edge of the clumps. The few nearly leafless stems, 30–60 cm tall, each bears a closed panicle 5–15 cm long. Spikelets each produce up to nine flowers, most of which set seed. The seed hulls are strongly awned.

Bluebunch fescue is one of the most palatable forages in the association where it grows. It has average protein and phosphorus contents, which are high during the spring but decrease as summer advances. Despite its lower contents of nutrients during the fall, its cured forage is sought by livestock all season and is eaten as readily in fall as it is during the spring. It is resistant to grazing and tramping, but when abuse reduces its stand it is replaced by species of needle grass or the less palatable timber oat grass.

A closely related species, *F. ovina* L., sheep fescue, grows in the same general area. Its leaves are gray green instead of blue green and are much shorter, and its awns are shorter than those of bluebunch fescue. Also, it is much less palatable and does not cure well, and is therefore less desirable as fall or winter forage.

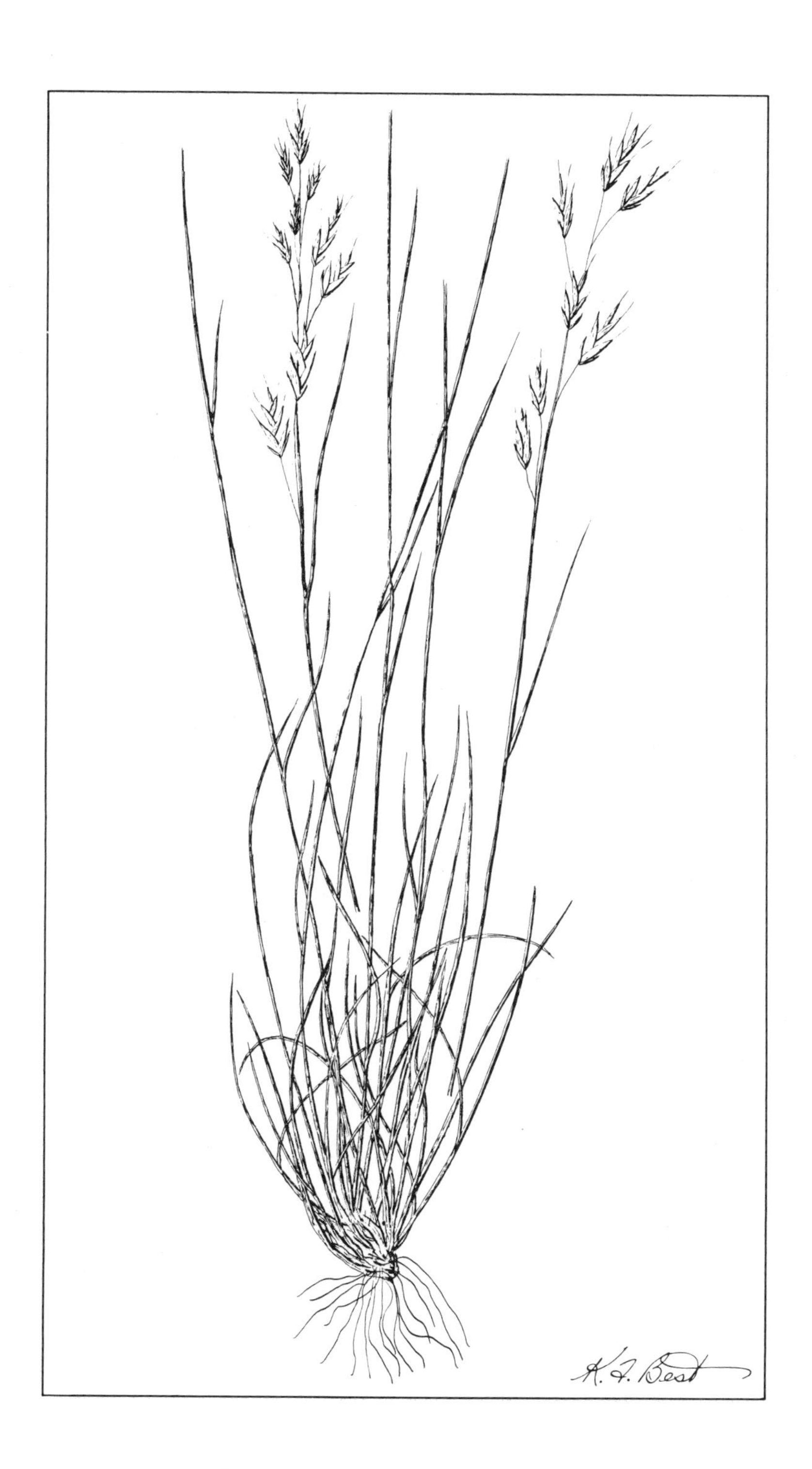
K. F. Best

Creeping red fescue *Festuca rubra* L.

Creeping red fescue is a temperate zone plant with a wide distribution. Forms of the species are native to North America, North Africa, Eurasia, and Iceland. In North America strains of the species are found throughout the Allegheny Mountains and Atlantic marsh areas, as well as in certain coastal ranges of the Rocky Mountain formation. It has proved to be useful as a pasture grass at the western and eastern edges of the prairie area, in mixtures for irrigated pastures in southern Alberta, and for prevention of erosion on irrigation ditches throughout southern Saskatchewan. It grows farther north than any other cultivated grass.

Creeping red fescue has variable creeping habits. Some strains spread only by tufts from the central crown, whereas others have fairly strong rhizomes. It has deep feeding roots. All leafage is basal, shiny, folded, and bright green except for the reddish lower sheath. Stems are nearly leafless, shiny, and up to 80 cm tall. The seed head is a closed panicle with purplish-tinged spikelets. Seed hulls are awned.

Creeping red fescue is extremely palatable at all seasons of the year. It maintains its green color into the autumn, as well as an above-average protein content. Ranchers in the Foothills regard it as the best grass for cultivated pasture in their area. It grows well by itself or in mixtures with white and alsike clover. It is essentially a pasture grass, as its leafage is too short to be cut and raked readily. In a test undertaken in central Manitoba, cattle preferred it to timothy and smooth brome during October and November.

Chewing's fescue, *F. rubra* var. *commutata* Gaud. (=*F. fallax* Thuill.), is closely related. However, it does not possess extensive creeping rootstocks and its leafage is a brighter green than that of creeping red fescue. It is used extensively in lawns, golf fairways, and pasture mixtures. It develops a dense turf that is resistant to grazing, cutting, and trampling.

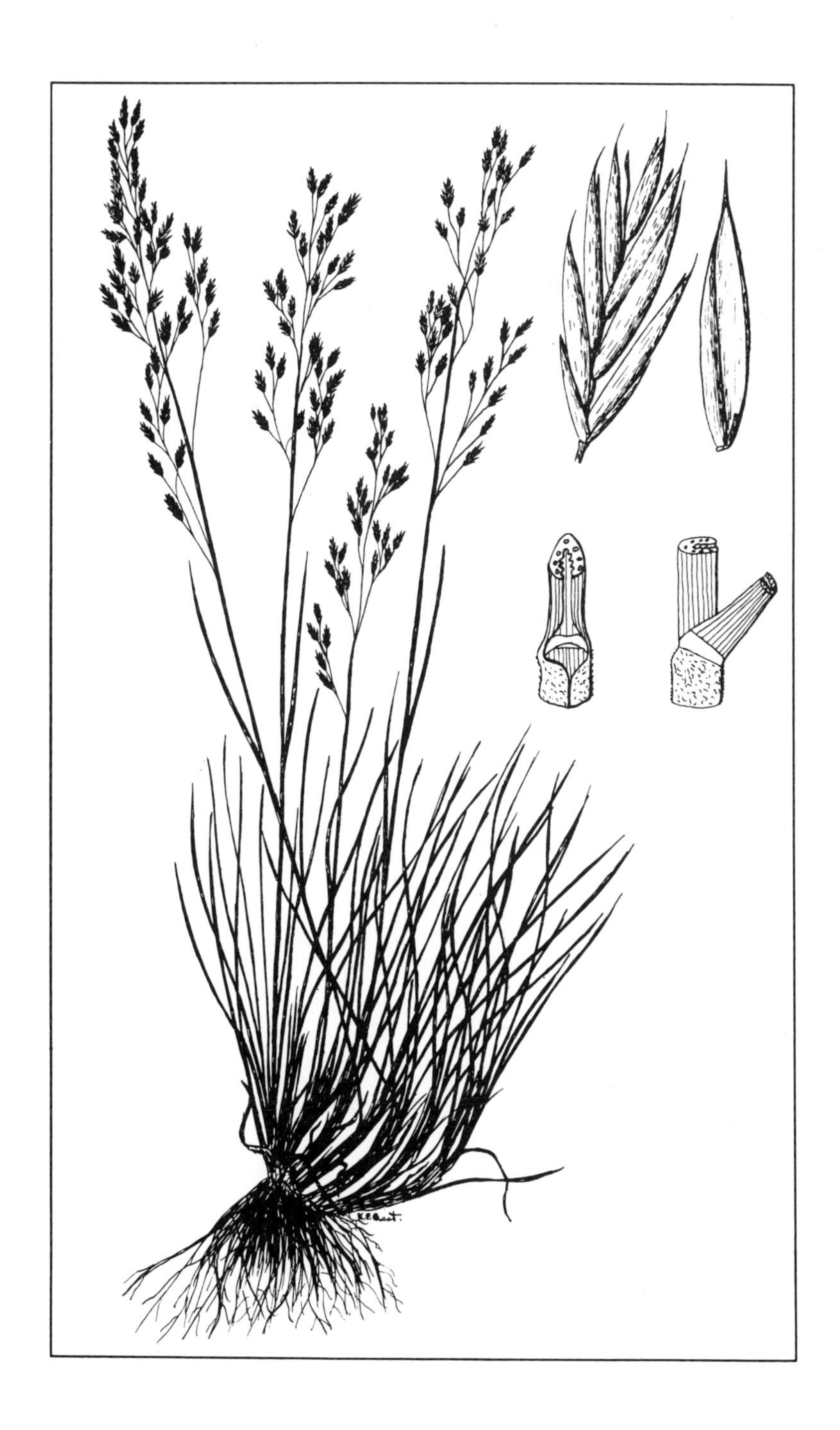

Plains rough fescue *Festuca hallii* (Vasey) Piper

Plains rough fescue is a tufted, native bunch grass. It occurs in the grasslands of the Cypress Hills, Wood Mountain, Hand Hills, Neutral Hills, and the parklands. It is seldom abundant where the annual precipitation is less than 400 mm and where the mean annual temperature is more than 2°C.

Densely tufted, gray green leaves produce the bulk of the forage. Growth commences early in the spring, and the seed matures in 90–95 days. Individual plants spread from short rhizomes. Plains rough fescue is palatable and is easily grazed out. Under heavy grazing its place is taken by northern porcupine grass, needle-and-thread, or weedy species, which may become dominant when overgrazing is continued.

Plains rough fescue cures on the stem, but early frosts may wither the plants before curing is completed. Spring protein content is low, and livestock gain slowly in the spring in pastures where plains rough fescue is dominant.

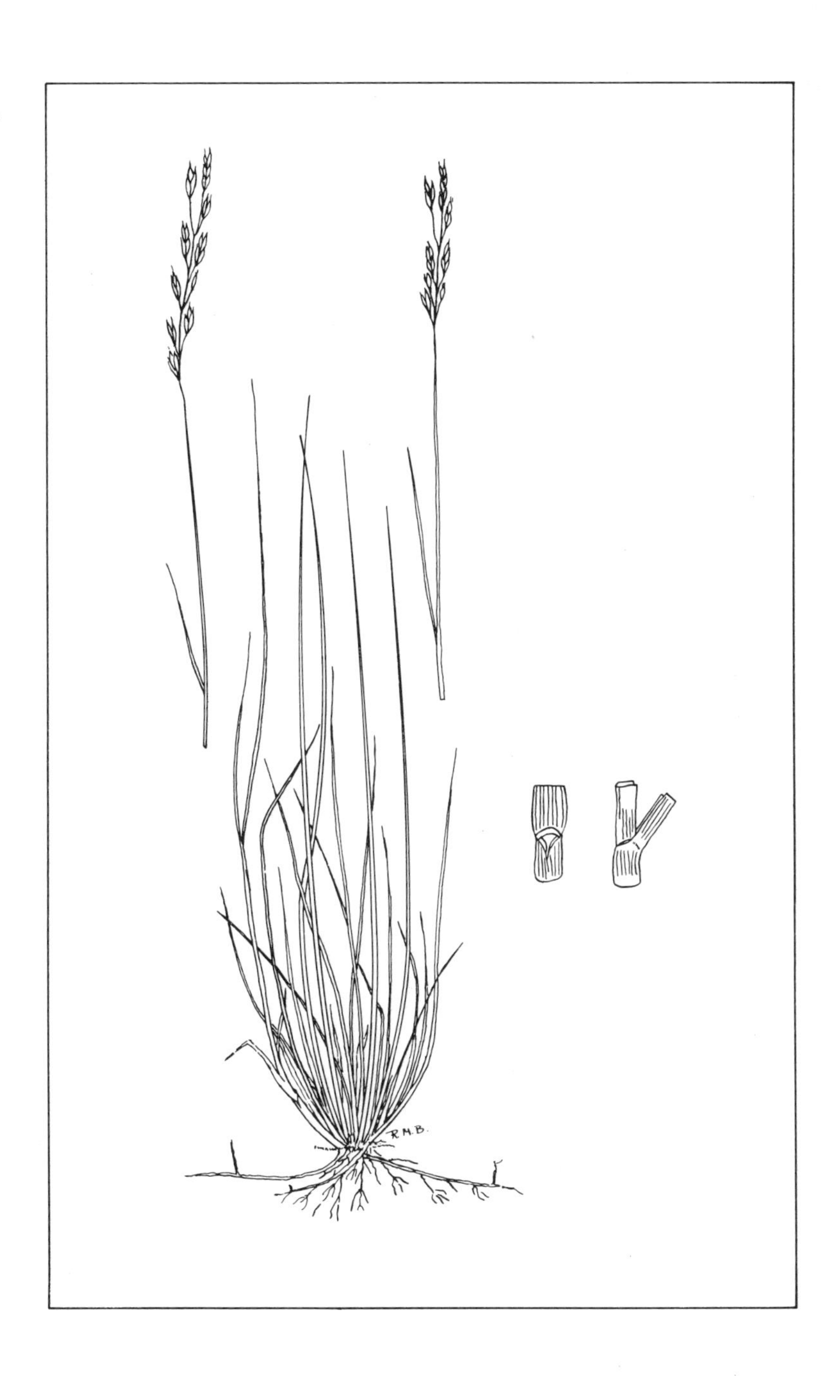
R.M.B.

Rough fescue *Festuca campestris* Rydb.

Rough fescue is a large-tufted, native bunch grass. It is found as individual plants or as dominant in grassy openings and open forest in the Foothills of the Rocky Mountains as far north as the James River.

Dense, gray green, rough basal leaves produce a bulk of forage. The tall, leafless stems produce easily shed seed. Growth commences early in the spring and seed matures within 90–95 days. Individual plants spread from tufts growing at the edge of the crown.

Rough fescue is palatable and is grazed out fairly readily. As the grazing load increases, timber oat grass, shrubby cinquefoil, pasture sage, and other lower-producing grasses or weeds become dominant. Management that allows for a complete rest or moderate grazing and haying in a 6-year rotation apparently maintains stands in a productive condition. Rough fescue cures on the stem, but early frosts sometimes wither the plants before curing is completed. It makes good hay, with a dry matter digestibility of over 55% when cut before the end of July.

The rough fescue that grows in the Cypress Hills has a peculiar protein cycle. Instead of the 16% crude protein content possessed by most native grasses during May, rough fescue does not have more than 12% protein. This value increases slightly during June, but by August the protein content drops to about 7% and is seldom more than 4.5% by mid-October. Crude fiber and lignin contents are low, but digestible carbohydrates are high at all times. The low protein content in spring may cause the slow livestock gains in spring observed in certain districts where rough fescue is dominant. The forage value of rough fescue appears to vary from district to district; rough fescue from the Foothills seems to have higher nutritive ratings that rough fescue from the Cypress Hills.

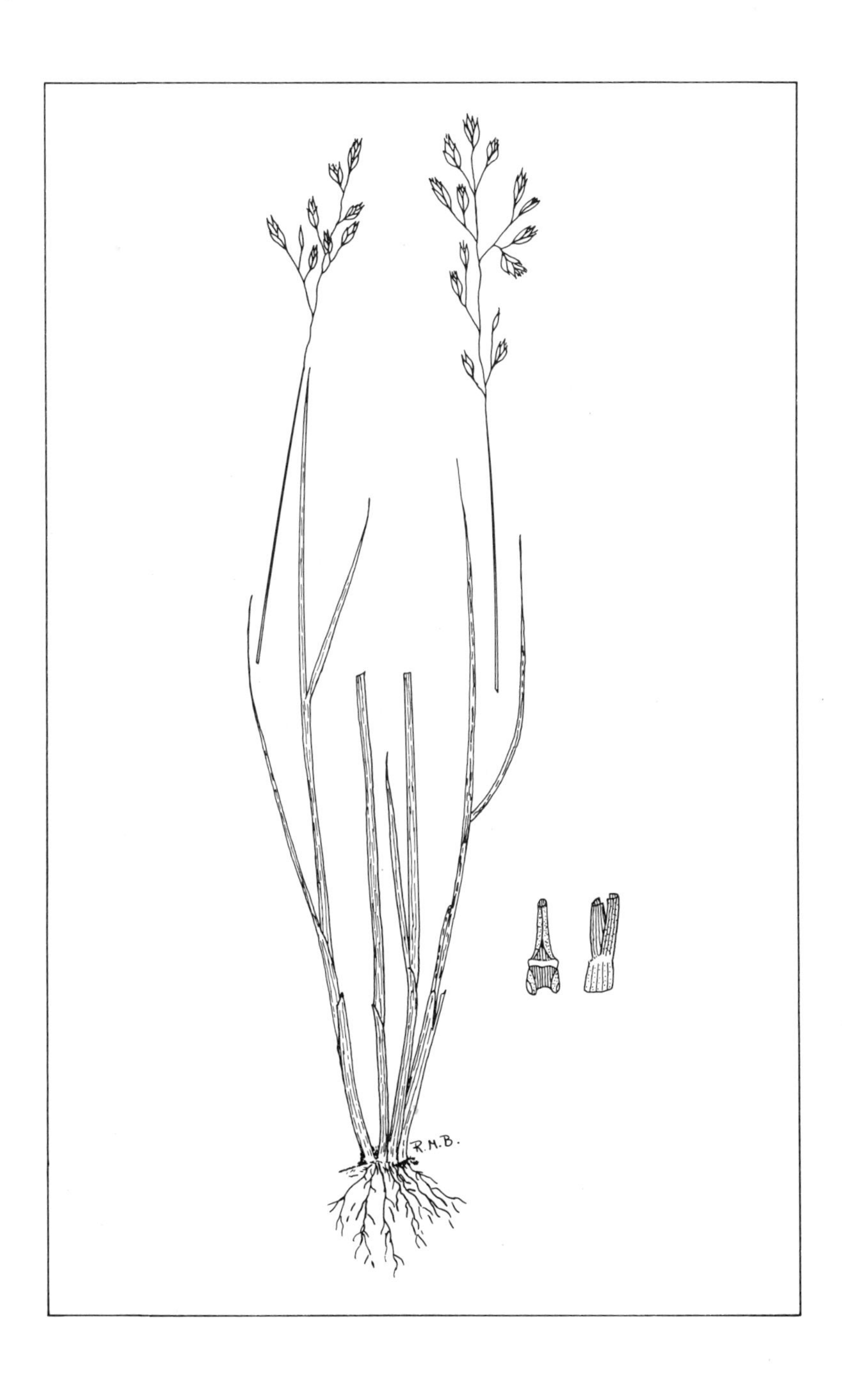
R.H.B.

Tall fescue

Festuca elatior var. *arundinacea* (Schreb.) Wimm.

Tall fescue was introduced into Canada from England, but it is a native of Central Europe. Since its introduction, it has been cultivated for pasture and hay for several years and has spread from fields into waste lands in certain districts. It appears to be adapted to the moist conditions of Eastern Canada and the United States, as well as to favorable sites in the western half of North America.

Tall fescue has an open bunch habit of growth. Plants extend their basal area by tufts that grow from the edge of the crown. Coarse, leafy stems 1 m tall support an open, nodding panicle. Seeds are numerous and short awned.

Tall fescue has not been grown to any extent in the prairie area. Improved strains have been selected for marshy and wet soils in Eastern Canada and work is being conducted to determine its place elsewhere.

A closely related and very similar species, meadow fescue (*Festuca elatior* L.), has been tested and does well in mixtures with smooth brome and timothy at the edge of the prairie area. It develops stands quickly and has some drought tolerance, but it is shorter lived and less palatable than timothy, smooth brome, or alfalfa. It has been used to some extent in irrigated pasture mixtures in southern Alberta and Saskatchewan.

Foxtail barley *Hordeum jubatum* L.

Foxtail barley, or wild barley, is a short-lived perennial closely related to the common cultivated barley. It grows throughout North America in abandoned fields, saline marshes, roadside sloughs, and waste places. It may become a weed in irrigated pastures and hay fields.

Foxtail barley has a dense, shallow root. It produces an abundance of bright green basal leafage and many seed stems that become straw colored at maturity. The spike has three spikelets at each node, each containing a single seed. Spreading, barbed awns 2–5 cm long grow from the tip of the seed, and equally long, barbed glumes grow from the base of the seed. The seed head, or spike, breaks up at maturity and is carried by wind, clothing, sheep's wool, and machinery. The plant depends entirely on its bounteous seed supply to maintain its stand.

Foxtail barley is a palatable forage until the heads form in the sheath. Prior to this date, usually in early June, foxtail barley produces palatable pasture that has high protein and phosphorus contents and a low crude fiber rating. Around saline sloughs and marshes it grows as the water recedes, and this habit extends the grazing season wherever such conditions occur. It also produces nutritive, though low-yielding, hay. However, its long, barbed awns make it a dangerous hay, because they catch in the gums and other parts of the mouth where they fester and cause smelly and unsightly abscesses. Many farmers and ranchers use foxtail barley hay regularly during the winter by feeding it in the open on top of snow where the animals have an opportunity to select what they desire, and the snow in contact with the heads makes them softer and thus more palatable. Cattle can make better use of foxtail barley than sheep or horses.

K. F. Best

Indian grass *Sorghastrum nutans* (L.) Nash

Indian grass species are generally warm-weather grasses growing in the eastern half of North America. Only one species occurs in the prairie area, and its range is within the drainage basin of the Red River in Manitoba. The generic name *Sorghastrum* means resembling a sorghum. The name is quite appropriate because plants resemble tall spindly sorghums, and their feeding values are relatively low after maturity.

Indian grass is one of the common grasses composing the cover in the area adjacent to the Red River in Manitoba. The associated grasses include big and little bluestems and switch grass.

Indian grass is a creeping-rooted species. It also has a dense, deep root system for feeding itself. Leaves are relatively few, flat, and tapering to a point, growing as long as 40 cm. The leafy stems may be as tall as 2 m, but they average about 1 m in height. Seed heads or panicles are 15–30 cm long, quite bushy, dense, and usually yellow. A large quantity of seed is produced. The 10-mm awns that extend from the seed hulls give a bristly appearance to the panicle.

Indian grass is palatable in its early stages of growth. However, as with most other coarse grasses, it loses its tastiness and feeding qualities after heading occurs. Protein content drops from 18% in early spring to 6–7% by August, whereas fiber increases from about 25 to 35% during the same period. Frost destroys its nutritive qualities. It does not cure on the stem.

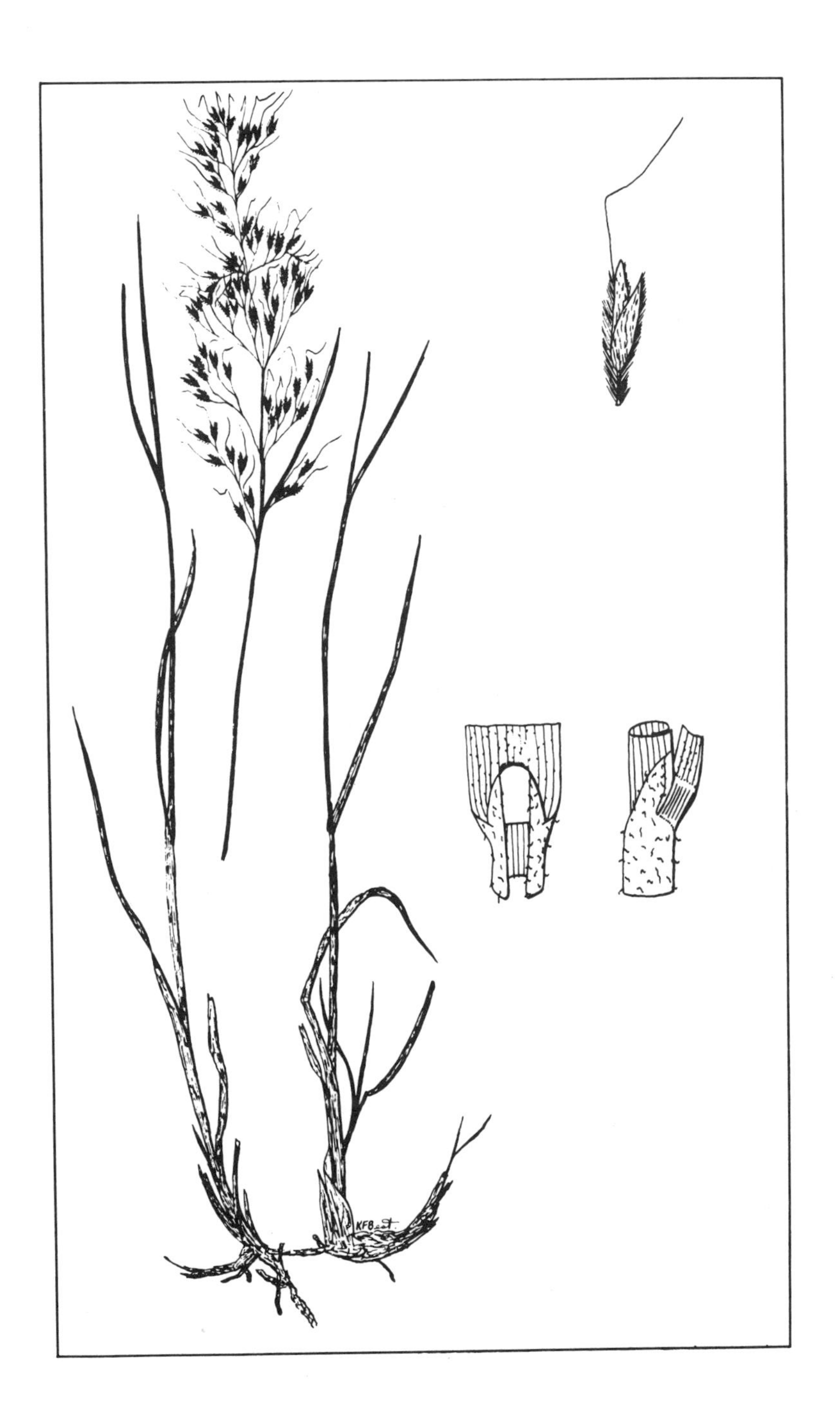
KFB

Indian rice grass

Oryzopsis hymenoides
(Roem. & Schult.) Ricker

Rice grasses have been named because certain species supposedly resemble cultivated rice. There are 12 species of rice grass growing in North America. Four of these occur in the prairie area, but only one, Indian rice grass, is important as a range forage.

Indian rice grass occurs throughout the western half of North America from central Alberta to southern Mexico. In the days of early settlement in Colorado and Utah, it was very abundant and highly regarded as a range grass. Indians ground its plump seed into flour from which they made bread. In the prairie area it is common on sandy soils and abundant locally throughout the Great Sand Hills.

Indian rice grass is a bunch grass growing from deep feeding roots. Its old leaf sheaths persist at its base and help protect the numerous long, deep green, folded basal leaves. Leafless seed stalks grow from a tuft of leaves to a height of 60 cm. The feathery panicle has a whitish color caused by the light-colored chaff. There is a single, round seed in each floret. The seeds are surrounded with white hairs, are black at maturity, and are each tipped with a short awn.

Indian rice grass is rated one of the most palatable of native grasses. Undoubtedly its palatability has contributed to its disappearance. Its green spring growth and lush summer foliage seem out of place on the sand dunes and dry sandy soils where it is most abundant today. It cures well in early autumn to provide nutritious and palatable winter grazing. Its hard seed coat reduces its germination for at least a year after maturity and the hairs that surround the seed have to be processed before it can be seeded through a drill. Its protein content reaches 10–12% in spring and early summer but declines rapidly at maturity to about 7%, a level that is maintained until early autumn.

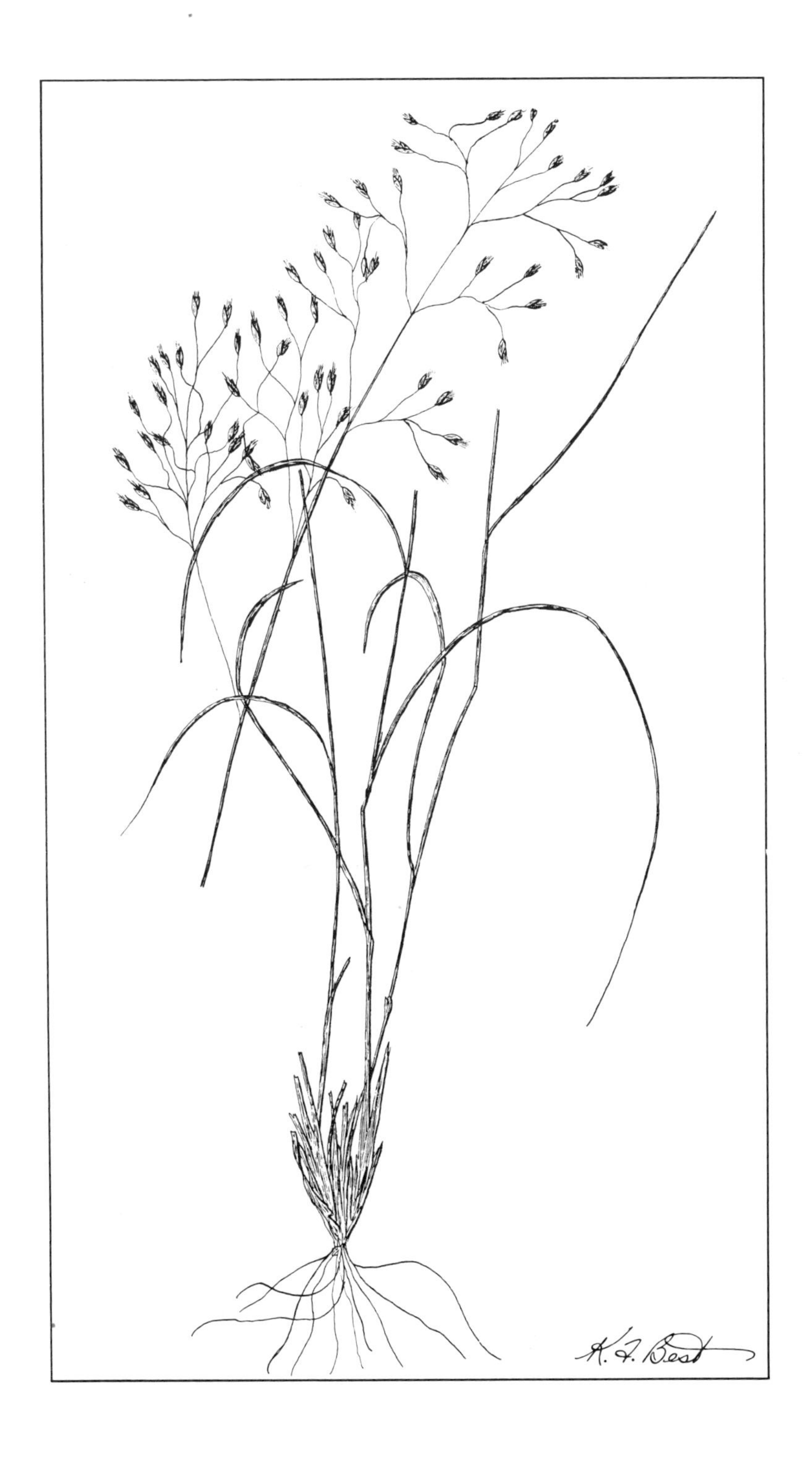
K. F. Best

June grass *Koeleria macrantha* (Ledeb.) Schult.

June grass is a long-lived bunch grass native to North America, Eurasia, and Africa. It is found throughout the prairie area. It has long been known as *K. cristata* (L.) Pers. or *K. gracilis* Pers. In the short-grass prairie it grows in association with blue grama and needle-and-thread, in the Foothills with rough fescue, in open glades in the northern boreal forest with hairy wild rye grass and reed grasses, and in the interlake area in Manitoba with little bluestem.

June grass seldom grows in dense stands, but rather it occurs as single plants in mixed communities. It is shallowly but densely rooted to a depth of 10–20 cm, with a few deeper roots extending 60–120 cm. Short, ribbed, flat, dark green leaves grow in tufts, and slender, erect, nearly leafless stems grow to a height of 10–60 cm. The head is a closed or tight panicle, almost like a spike or column, that opens during the short flowering season in mid-June. Seed is abundant, usually of low fertility, and matures within 75–80 days after growth commences in the spring. Plants vary considerably in their appearance under various growth conditions, particularly in height of stem, which increases as better growth conditions occur; the short, ribbed basal leaves are characteristic.

June grass is palatable during the spring and late fall. However, it appears to lose its palatability after the flowering period and until curing is completed. In has a 20% protein content during the early spring, but this percentage decreases rapidly during June and July to a low of 4% by November. Phosphorus, crude fiber, and digestible carbohydrate contents are similar to those of the good prairie grasses. June grass cures on the stem. Although June grass is one of the most widespread and common range grasses, it is seldom very abundant and of little economic importance.

Mat muhly *Muhlenbergia richardsonis* (Trin.) Rydb.

Over 50 species of muhly grow in North America. The home of most is along the Atlantic seaboard and the Gulf of Mexico. Four species occur in the prairie area, but only one, mat muhly, is abundant and widespread. Another species, scratch grass, *M. asperifolia* (Nees & Mey.) Parodi, occurs commonly in fairly dense stands in slightly saline areas, particularly in sandhill areas. In general, muhlys are rated as from poor to fairly good forage plants.

Mat muhly is a perennial that occurs throughout southern Canada and west of the Mississippi River in the United States. In the prairie area it is found on shallow, moist, slightly saline soils, as well as on dry upland from the interlake area in Manitoba to elevations of 1800 m in the Rocky Mountains. It invades eroded sites where better forage grasses cannot establish themselves.

Mat muhly has a dense, shallow root system, as well as fine creeping rootstocks. Dense tufts of short, dull green leaves surround the one to many seed stems. The stems grow in a zigzag fashion from node to node and are each topped with a slender, interrupted, closed panicle. Very little seed is produced.

Mat muhly is not very palatable. Its wiry leaves and stem are probably hard to bite off; they certainly are difficult to mow. However, its use is undoubtedly dependent on the abundance of associated species, because it is eaten when more palatable species are scarce. Mat muhly cures on the stem and is eaten more readily after the first snowfall than in midsummer. Protein and phosphorus contents are somewhat lower in the spring than are those of common prairie species.

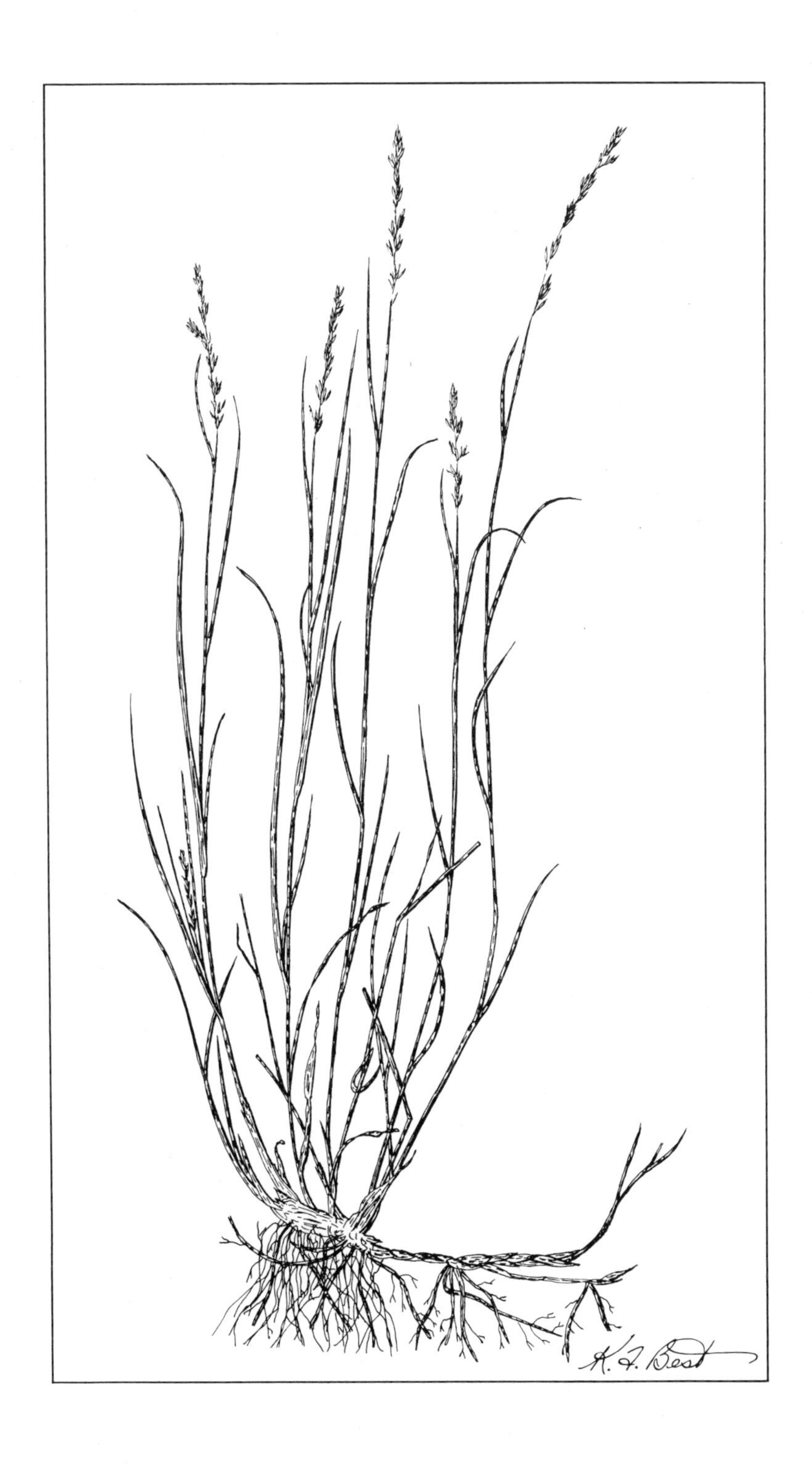
K. F. Best

Green needle grass *Stipa viridula* Trin.

Needle grasses have a worldwide distribution. Most are excellent forages, but a few have sharp-pointed seeds that cause mechanical injuries to livestock and one is known to be poisonous. Six needle grasses occur in the prairie area.

Green needle grass is known also as green feather grass. It is a grass of the central Interior Plains of North America, although there are a few records of its occurrence near the Atlantic seaboard. It is found everywhere in the prairie area, although few dense stands occur. Its best growth is made on heavy clay soils where it grows in association with western wheat grass. Good growth is made also on sandy land if its deep feeding roots can reach a water table.

Green needle grass is a bunch grass with a dense root system that feeds down to a depth of 2–3 m. Long, green, narrow basal leaves surround the 1.25-m greenish stems. Small, black, hair-covered seeds are produced in abundance. There is no sharp pointed callus as in other needle grasses and the awn is weak and seldom over 5 cm long. There are white hairs at the junction of the leaf blade and sheath and along the edges of the sheath.

Green needle grass is regarded as the most palatable of the needle grasses. It starts growth in late April, matures seed by mid-August, and cures by mid-September. During these several stages of growth all livestock seek this species and graze it readily. Spring protein content is over 20%, and although some reduction of protein occurs as the plant matures, even late fall protein content is often over 8%. The ability to retain a relatively high protein content is one of the outstanding characters of this grass. In addition, it is tolerant to drought and cold and moderately resistant to grazing.

A cultivar of this species has been improved and named green stipa grass. It has been used in comparative trials, but its relatively low productivity and short useful life make it inferior to most of the introduced forage grasses.

A very similar species, *Stipa columbiana* Macoun, occurs in the montane grasslands, Cypress Hills, and Peace River regions.

K. F. Best

Needle-and-thread *Stipa comata* Trin. & Rupr.

Needle-and-thread, or common spear grass, is a long-lived perennial that occurs throughout Western Canada and extends southward through the United States into Mexico. It is one of the most important native grasses in the prairie area, although its stands become progressively sparser from the dry central portion, where it produces nearly 50% of the pasturage, to the parkland, where it is quite rare.

Needle-and-thread is a bunch grass with a root system extending about 1.5 m beneath the surface. Its gray green leaves grow in a dense cluster. One to ten straw-colored stems are prostrate at the base and grow to a height of 60 cm. The panicles are loosely spreading and branched. Each branch carries several one-flowered spikelets. The seed hull is round, straw-colored, and about 12 mm long, ending in a twisted, bent awn 20–25 mm long. A sharp, needlepoint callus occurs at the lower end of the seed.

Needle-and-thread is probably the most palatable of all native grasses. Its leaves start growth in late April and grow rapidly until mid-June, at which season the shot blade emerges. Flowering commences in early July and seed matures and drops by early September. Protein content of spring growth is over 20% but decreases to about 5% when curing is completed. The species is drought tolerant and resistant to frost, but it can be grazed out with heavy pasturing.

Recent studies have shown that the leaves are light in comparison to their length during the spring. Therefore, needle-and-thread pastures should be protected until mid-June. Other work has shown that a cow requires about 5000 plants a day to satisfy her appetite.

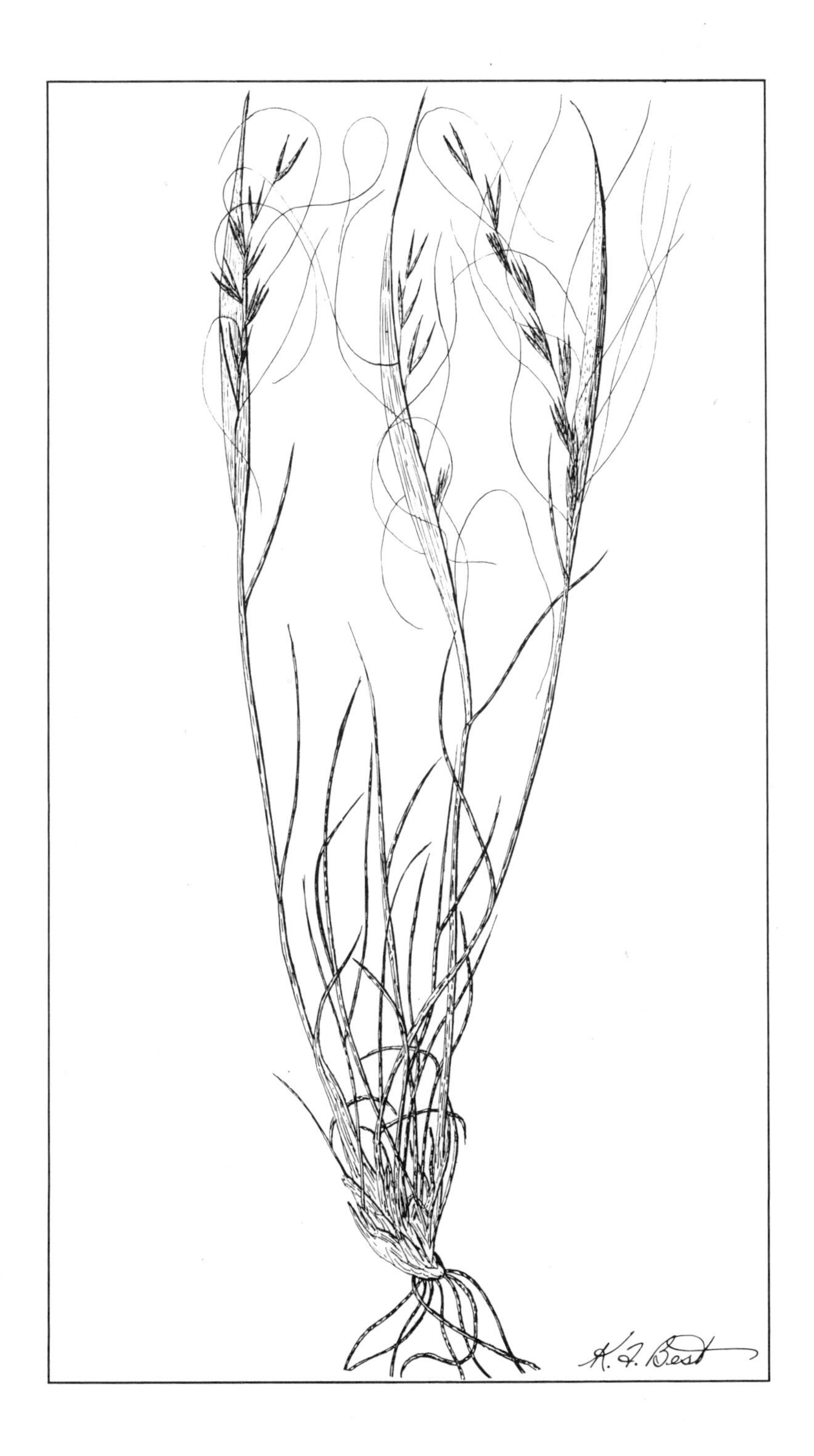
K. F. Best

Northern porcupine grass

Stipa curtiseta (Hitchc.) Barkworth

Northern, or western, porcupine grass is not a particularly good name for this species because it looks more like needle-and-thread than it resembles the robust and harsh porcupine grass. In fact, its leafage is hard to distinguish from needle-and-thread, although its leaves are somewhat longer and greener, pale green instead of the gray green of needle-and-thread.

Northern porcupine grass is usually the dominant species on loam soils in the central part of the prairie area. Whereas needle-and-thread is the dominant needle grass in the dry central area and porcupine grass may be the dominant needle grass in the parkland, northern porcupine grass is the most abundant needle grass in the area between these two extremes. It is found only in the Prairie Provinces and the states of Montana, Wyoming, and the Dakotas.

Northern porcupine grass is a short-leaved bunch grass. Its roots extend to a depth of 1 m. The flowering stems are about 1 m tall and produce up to 30 sharp-pointed seeds in the feathery panicles. The seeds are about 15 mm long, each terminating in a 10-cm awn that is twisted and bent. In undisturbed stands the old basal leaf sheaths cling to the crown, as do those of porcupine grass.

Northern porcupine grass is fairly palatable to all classes of livestock except during the period between emergence and dropping of the seed. This grass starts growth late in April, matures and drops its seed by mid-August, and cures by September. In general its nutritive and curing properties are considered to be somewhat poorer than those of needle-and-thread, but these properties are rated much higher than those of porcupine grass. Its calcium-to-phosphorus ratio after curing is approximately 5:1, about the same as for all good native grasses.

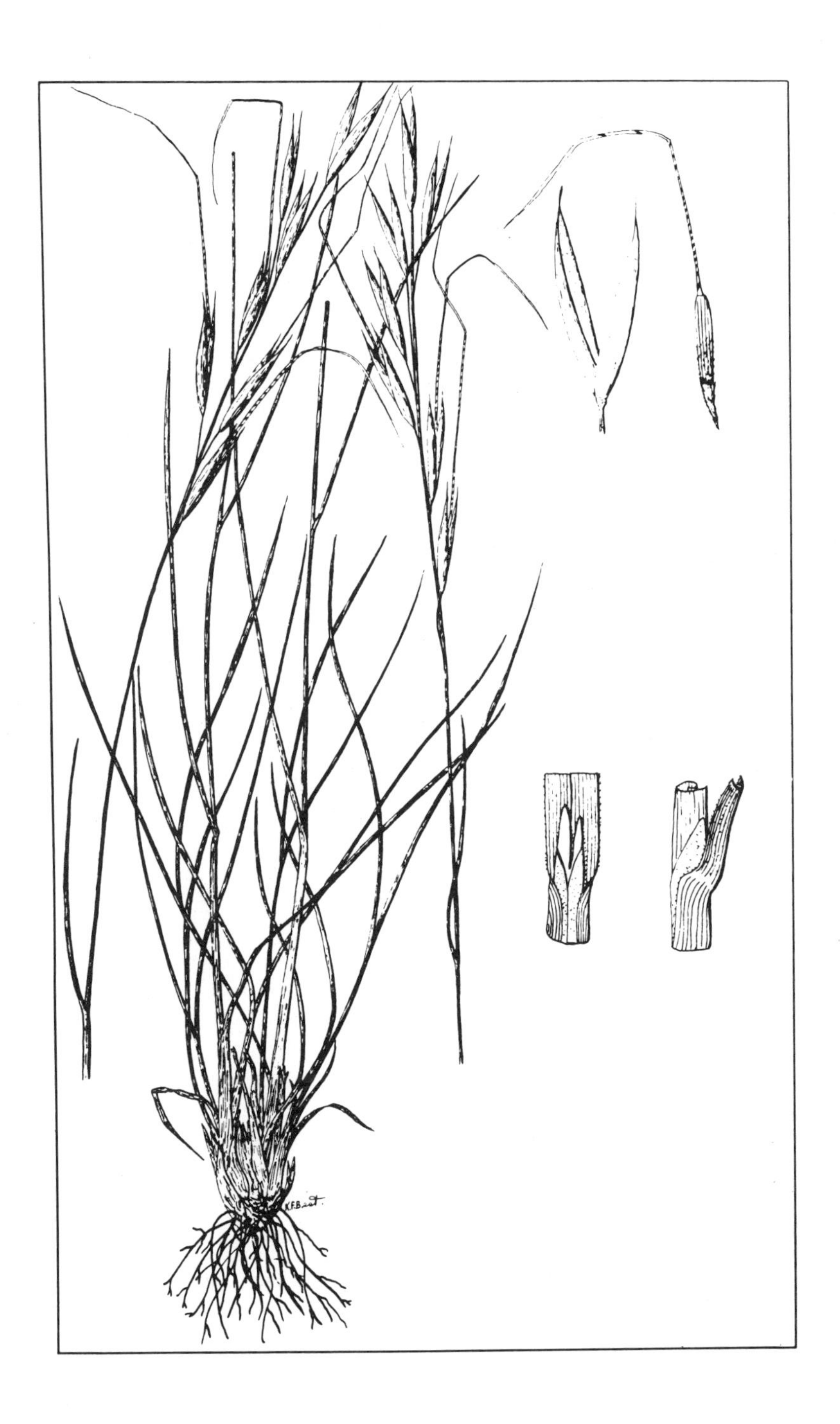

Porcupine grass *Stipa spartea* Trin.

Porcupine grass is an eastern species, locally abundant in Eastern Canada and the northeastern United States. Within the prairie area it is fairly common throughout western Manitoba, with stands occurring on sandy soils adjacent to the Assiniboine River in both Saskatchewan and Manitoba. Plants have been collected also throughout the parkland region of the Prairie Provinces.

Porcupine grass is a robust bunch grass. Deep feeding roots form a dense mat near the surface and thus effectively control erosion on the sandy plains where the species makes its best growth. Many narrow leaves over 30 cm long grow to form a very coarse forage. The whitish stems, which may be over 1.5 m tall, each terminate in an open panicle that produces up to 30 or more seeds. The seeds are nearly 25 mm long, including the needle-sharp tip. A long, twisted, twice-bent awn grows from the upper end of the seed. Seeds catch on clothing, hair, fur, and wool, after which the sharp callus easily penetrates the skin. Another distinctive character of the species is the persistence of the old basal leaf sheaths, which form a very straggly crown.

Porcupine grass is poor feed. Even its early spring growth is coarse and not too palatable. From the time the seed head emerges from the stem until the seed is dropped, no animal willingly ventures through a stand. After the seed drops off, some grazing may be obtained. In the few districts where porcupine grass is dominant, a common management practice is to burn off the old growth, a treatment that does not injure old plants but does prevent the establishment of seedlings. The prairie area is fortunate in that there are very few large stands of this grass.

Nuttall's salt-meadow grass

Puccinellia nuttalliana (Schult.) A.S. Hitchc.

Nuttall's salt-meadow grass has a range that includes the western half of North America, as well as the states of New York and Maine and the Maritime Provinces of Canada. It is quite common throughout the prairie area, where it grows in saline soils with high water tables. Often it is the dominant species in this habitat type, and occasionally it forms almost pure stands.

Nuttall's salt-meadow grass is a perennial bunch grass with a relatively shallow root system. Numerous fine, flat, short basal leaves grow from the crown. From one to ten stems grow to a height of 60 cm. The very open pyramidal panicle produces a considerable seed crop with four to six seeds in each spikelet. The seeds are small and overlap each other in the spikelets.

Nuttall's salt-meadow grass is quite palatable in its early growth stages. As maturity advances palatability declines. In some districts this grass is harvested for hay, but its low yield in most areas makes it an expensive feed. Further, it has a relatively low protein content at maturity, 10–11%, and is high in ash and fiber. In permanently moist soils it often grows in association with the poisonous seaside arrow-grass, *Triglochin maritima* L., and heavy pasturing may result in livestock losses.

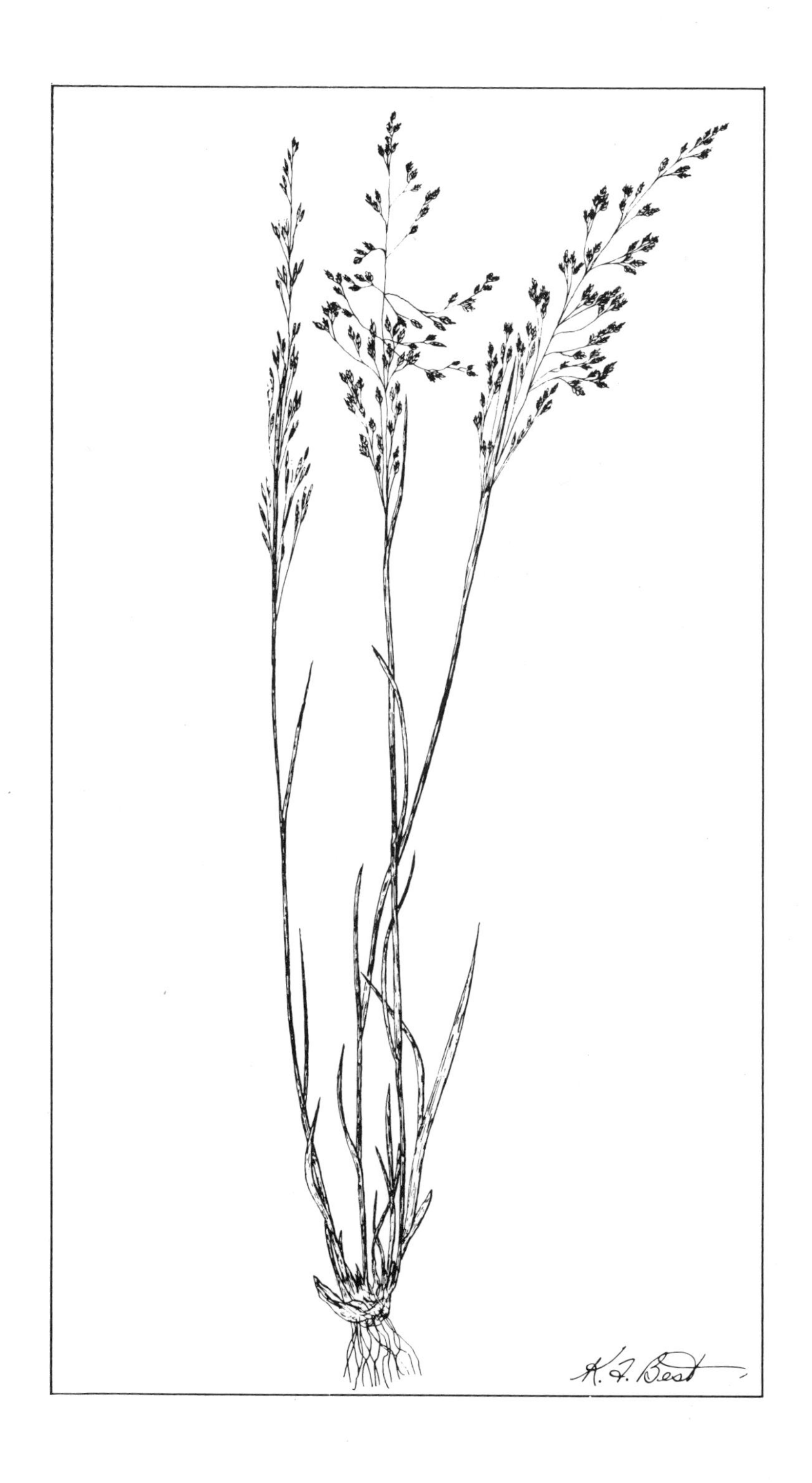
K. F. Best

Hooker's oat grass

Helictotrichon hookeri (Scribn. ex Hack.) Henrard

Only one species of *Helictotrichon* occurs in Canada, but in the Eurasian steppes the genus has numerous representatives.

Hooker's oat grass occurs only on the eastern slopes of the Rocky Mountains, from Alberta to Manitoba, and southward to New Mexico. Further, it grows only where moisture is fairly plentiful as in the Foothills, Cypress Hills, Handhills, Neutral Hills, and parkbelt of Western Canada.

Hooker's oat grass is densely bunched, with a shallow root system and flat leaves. It usually has one to several stems, each topped with a panicle whose erect branches cling closely to the central column. A spikelet consisting of three to six florets terminates each branch. Hulls of the seeds are firm, shiny, and brown.

Hooker's oat grass seldom occurs in solid stands. Rather, single plants are scattered among the associated species and seldom account for more than 5% of the stand. Thus it does not provide much forage and is not considered to be an important constituent of the rangeland cover. Its protein content, about 10% in mid-July, and the mineral content are equal to those of other range grasses, and observations indicate that it is palatable.

Parry oat grass *Danthonia parryi* Scribn.

The genus *Danthonia* has a worldwide distribution and is represented in temperate and subtropical rangeland of all continents. Five species occur in the prairie area, of which timber oat grass is the most common.

Parry oat grass occupies a restricted area in the southern Foothills. It has a narrow range east and west, about 50–70 km, and disappears entirely a short distance south of Rocky Mountain House in Alberta; it extends southward from Canada as far as New Mexico. Parry oat grass and rough fescue are codominants in the pastures of the Foothills, with Parry oat grass forming dense communities on slopes and ridges exposed to the chinook winds and rough fescue dominant in the sheltered locations.

Parry oat grass grows in tough clumps, the centers of which often die. Numerous overlapping, persistent basal sheaths enlarge the base and protect young growth. Leaf blades are light green or even yellowish green and somewhat hairy. Year-old blades drop off before growth commences in the spring. Short, erect stems are each topped with a closed panicle, which carries a few large spikelets. Seeds in the exposed spikelet are seldom developed, but those enclosed in the lower sheath are usually fertile.

Very few observations are available to indicate the palatability of Parry oat grass. It is so well equipped naturally to tolerate heavy grazing, trampling, and fire that it persists after less-resistant species disappear, although timber oat grass outlives it in pastures containing mixed stands. Where abundant, it is considered to be an indicator of good winter pasture. Chemical analyses suggest a good nutritive balance, with protein being lower during the spring in this species than in mixed prairie grasses but higher in midsummer and early autumn.

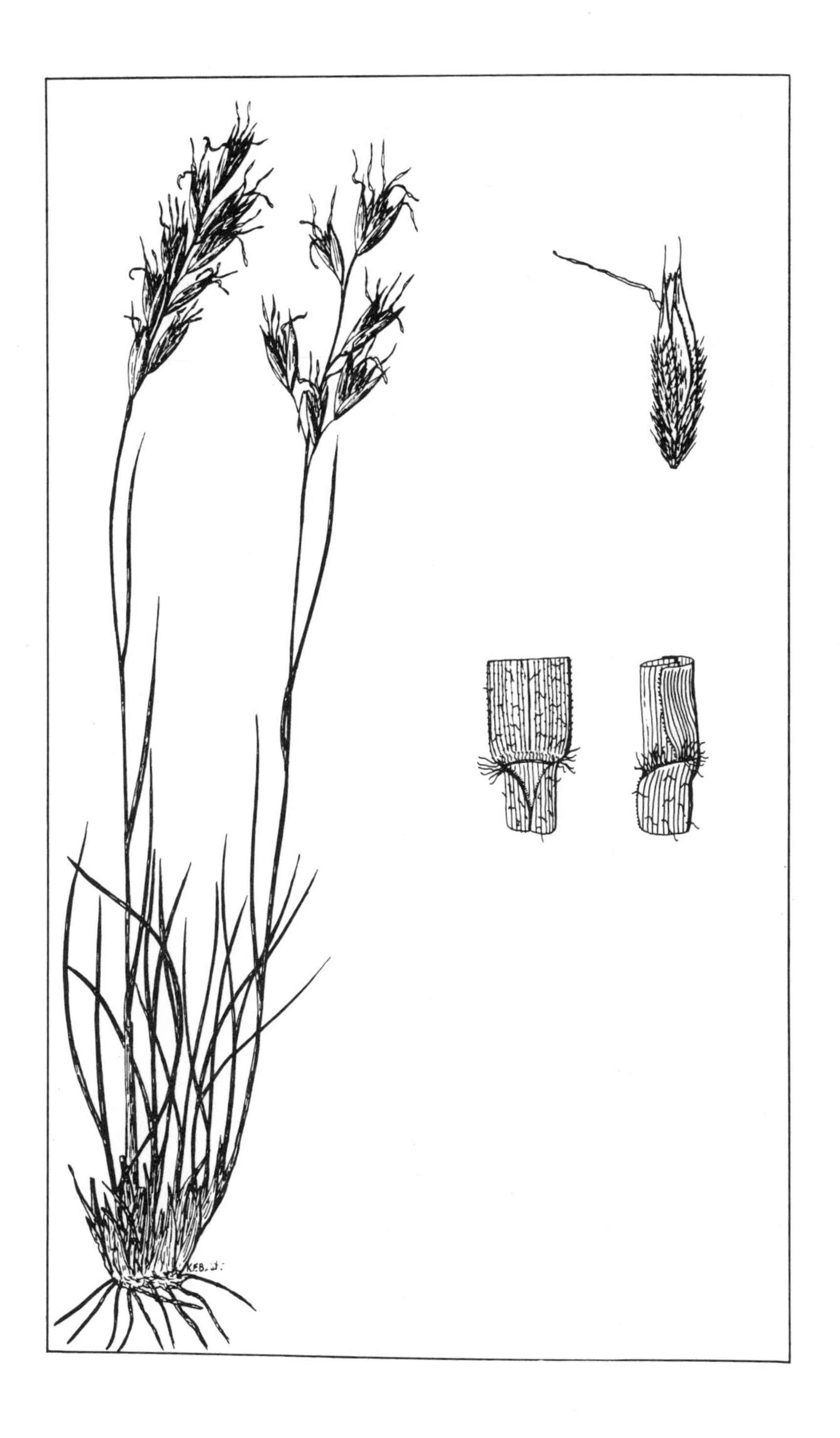

Purple oat grass *Schizachne purpurascens* (Torr.) Swallen

Purple oat grass is the only species in the genus *Schizachne*. This species has a range from coast to coast in Canada but is rarely found west of the Rockies in the United States. The name *Schizachne* is derived from the Greek and means split chaff, a reference to the split end of the seed husk.

Purple oat grass, also known as false melic, is a shade-loving grass. It grows only in forests and is more common in coniferous forest areas than in those where broad-leaved trees are dominant. Thus in the prairie area it is found only at the northern and western boundaries of the area and throughout the Cypress Hills.

Purple oat grass is a spindly bunch grass with a shallow, sparse, and rather coarse root system. The 1-m-high stem arises from a few flat, lax basal leaves. Usually a single leafless stem grows on one plant. The panicle is open and purplish in color, with a few, flowered, drooping branches usually occurring in pairs. A long awn extends beyond the seed hull.

Purple oat grass is not an important forage grass. It is seldom abundant, as it grows as single plants among the cover. Its yield is low and the grass itself is light, with a high water content when growing. No palatability rating has been assigned to this species, but observations suggest that it is only moderately palatable. Protein content is less than 10% and fiber content is high at all stages of growth.

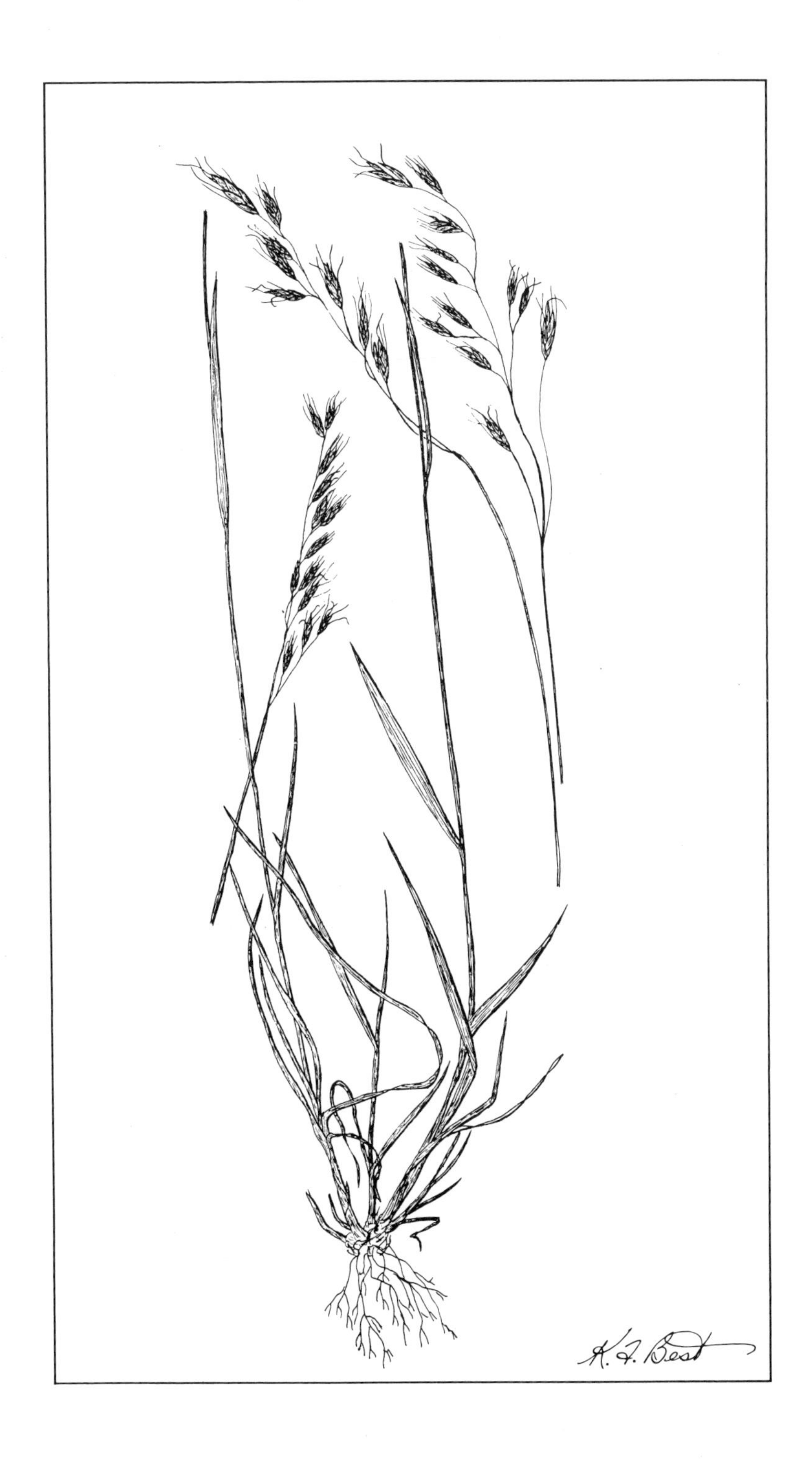
K. F. Best

Timber oat grass *Danthonia intermedia* Vasey

Timber oat grass is found from Quebec to British Columbia, and as far south as California and New Mexico. It grows on meadows and upland prairie close to the tree line, and at favorable sites within the spruce and pine zones. It is common in the Foothills and the Cypress Hills, and along the northern edge of the prairie area.

Timber oat grass usually grows in small clumps but on occasion it may form a dense sward. Plants produce an abundance of basal leaves and short but numerous stems. The four to twelve purplish spikelets that form the closed panicle are crowded on one side. Flowering occurs in August and the seed crop shatters early in September. Self-fertilized spikelets are hidden within the leaf sheath at the base of the stems, and after ripening the stems break off and distribute the enclosed seed.

Timber oat grass is doubtfully palatable. In the spring, when it grows rapidly, it appears to be eaten readily. However, in overgrazed fescue prairie it can form solid stands after the fescue has been grazed out, indicating that livestock prefers other grasses as long as they are available.

Chemical analyses indicate relatively low protein and phosphorus contents in all stages of growth, and a correspondingly high crude fiber content.

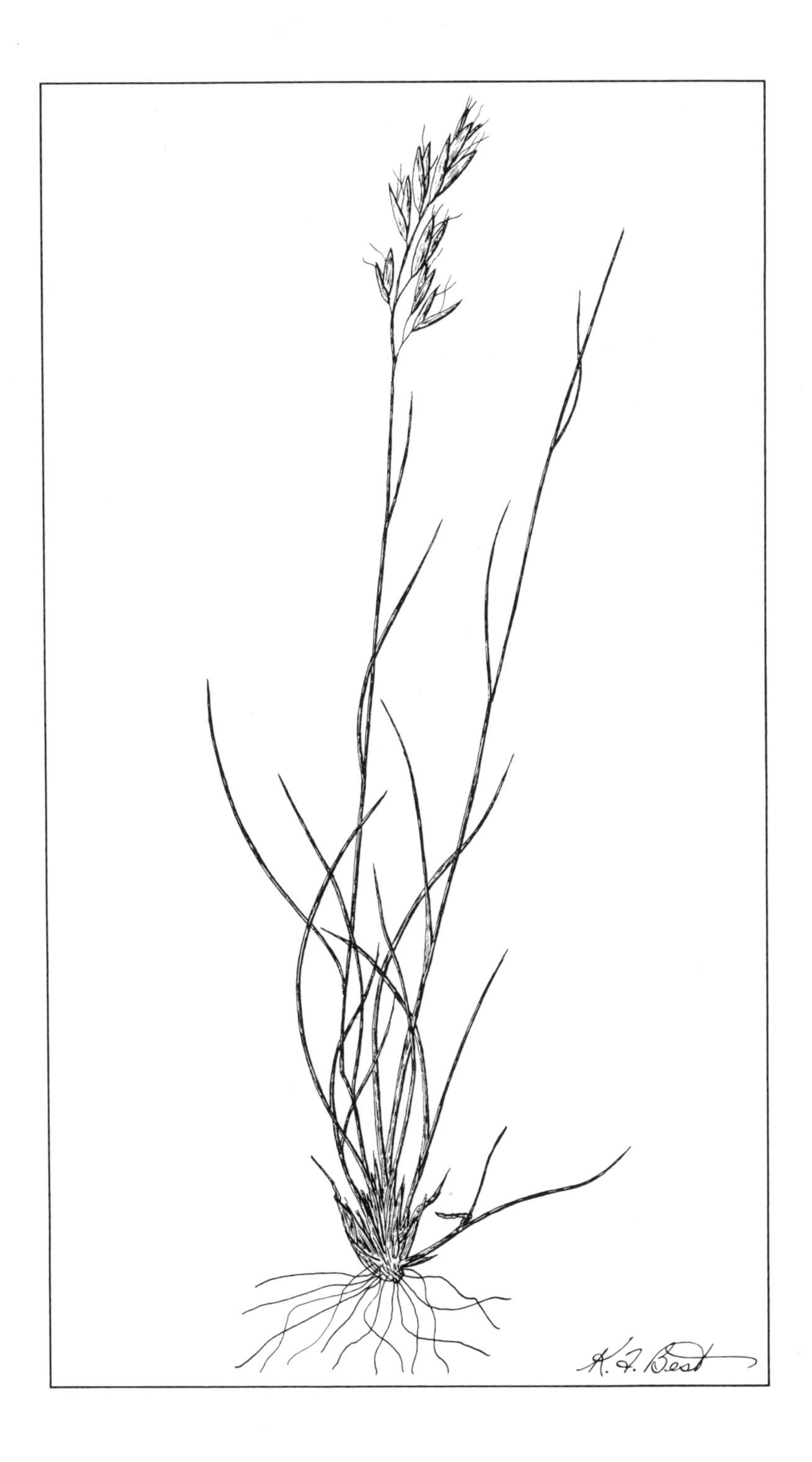
K. F. Best

Orchard grass *Dactylis glomerata* L.

Orchard grass is a native of Europe, where it is known as cock's-foot. It has been cultivated for many years and is now recognized as one of the best pasture grasses in countries with moist, temperate climates. It was first cultivated in North America about 1760 and has since spread throughout the continent. It is a useful cultivated pasture grass in Eastern Canada and has shown promise as a desirable crop for irrigated pasture in Western Canada.

Orchard grass is a bunch grass that forms large tussocks unless it is kept grazed down. It has a dense, deep root system, and produces long, folded, shiny, dark green leaves. The leafy stems are 1–1.5 m tall and are each topped with a flower head 10–25 cm long. In general the panicle is closed, but it opens during the short flowering period, at which time it has a purplish appearance. The panicle has a distinctive appearance. The flower clusters are not continuous; rather, there are breaks in the symmetry of the panicle.

Orchard grass is nutritious and palatable. It requires 500–650 mm of precipitation to produce high yields and nitrogen fertilizer to maintain high production. It grows well with alfalfa or red clover and, although more commonly used as pasture, it can be grown for hay. It is used in irrigated pasture as well as in orchards.

Reed canary grass *Phalaris arundinacea* L.

Reed canary grass is native in both North America and Europe. It was recognized as an excellent forage over 200 years ago in Sweden, and much of the seed used to establish stands elsewhere was imported from that country. It has been cultivated in Canada since early in the 1900s and is known to seed dealers as reed canarygrass.

Reed canary grass grows in clumps that spread by stout creeping rootstocks. Deep feeding roots provide nutrients for a dense cover of long, wide, light green basal leaves and tall, leafy stems. The head is a closed panicle, open only during flowering, and is interrupted. Only one seed is formed in each spikelet, but many spikelets grow on each branch of the head. The outer husk is whitish and shiny, and the enclosed seed is shiny brown and as long as a flax seed but narrower.

Reed canary grass is palatable. It produces good pasture that is eaten avidly and makes excellent hay either alone or supplemented with alfalfa. Although heavy seed crops are produced, the seed shatters easily making it difficult to harvest with farm equipment. As with all water-loving grasses, its hay is soft and somewhat wasty and is not the hard feed best suited for range cattle during cold weather. Protein content in good hay is 8–10% and its fiber content is about 37%.

Reed canary grass can stand ponding for 6–8 weeks, provided the water is shallow enough that the leaves can get above the surface and as long as the alkali content of the soil is low. It also grows on upland where more than 500 mm of precipitation occurs during its growing season or is provided by irrigation.

K. F. Best

Blue-joint *Calamagrostis canadensis* (Michx.) Beauvois

Reed grasses are distributed throughout the cool and temperate regions of the world. They are found growing in marshes, in the understory of forests, and in dry upland sites. Few are ever dominant over large areas, but instead are subdominants that fill in the cover.

Blue-joint, also known as marsh reed grass, is the most widespread and common of the several species, extending southward from the Arctic Circle to the southern states. It is a moisture-loving plant and grows in marshes, along streams, and in moist, shaded draws. Occasionally, nearly pure stands occur in shallow sloughs in the northern portion of the prairie area.

The creeping rootstocks of blue-joint give rise to many coarse stems that may grow to a height of 1.5 m. There is considerable basal leafage and the stems produce rough, lax leaves that may be 30–60 cm long. The nodding seed head is often 30 cm long, with the upright branches staying close to the central column. Seeds are numerous and small, less than 6 mm long, and are surrounded by a dense ring of hairs that arise from the base.

Palatability of blue-joint is not high at any season, although the leafage may be grazed in the spring. Palatability of this grass varies from district to district, and conflicting reports of its grazing characters have been published. Cut early in July, it is considered to be good hay. Excessive tramping reduces stands.

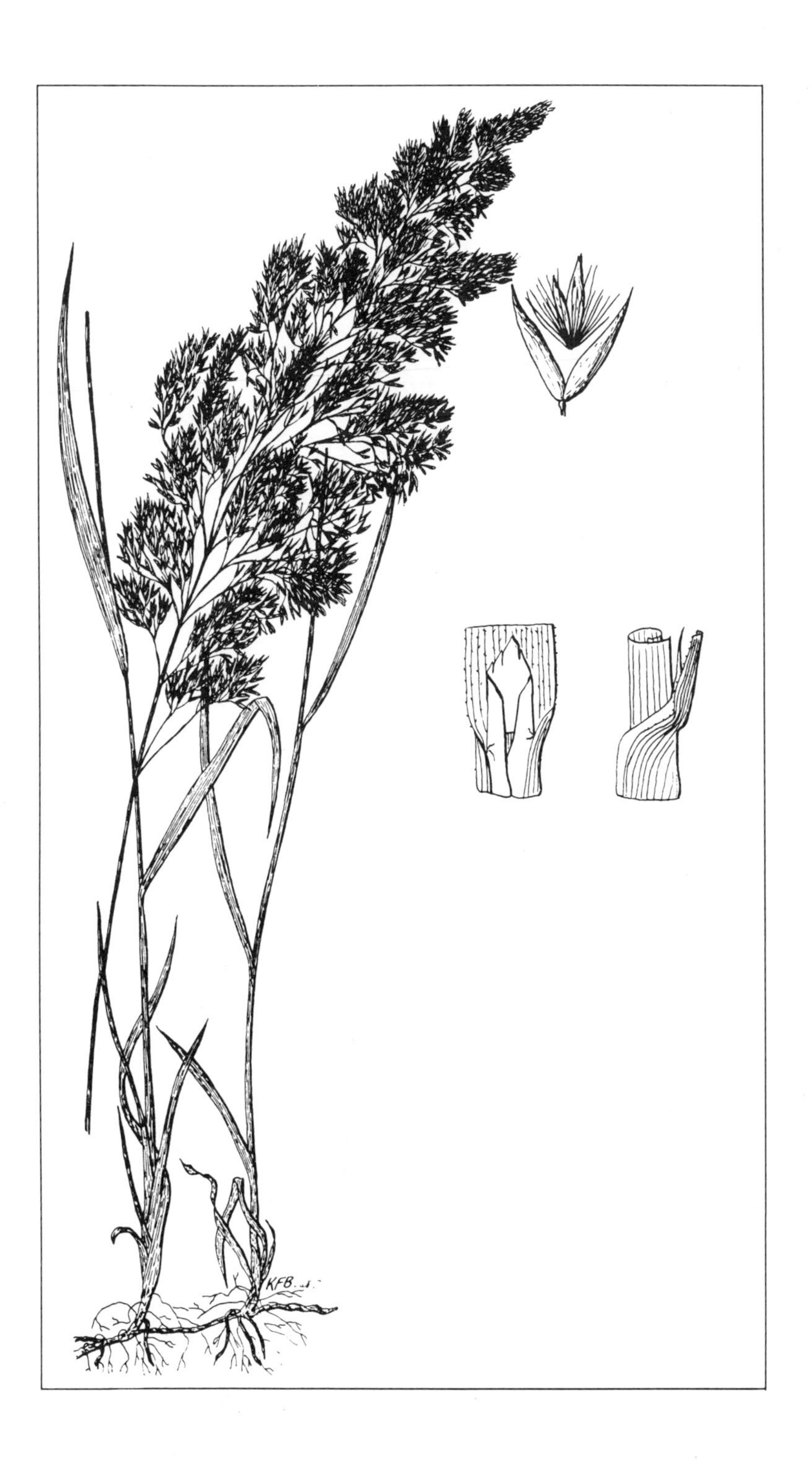

Northern reed grass *Calamagrostis inexpansa* A. Gray

Northern reed grass has a distribution similar to blue-joint but may be found at higher elevations in the Rocky Mountains. Its habitat includes springs, marshes, and meadows that have a fairly deep organic layer, and where the water table is close to the surface throughout the summer. It does not grow in heavy or extensive stands; rather, individual plants consisting of one to six stems and the basal leafage are found separated by other grasses and sedges.

Northern reed grass has stout creeping rootstocks. It produces a small amount of basal leafage and firm, rough stem leaves. Many leafy stems arise from the creeping root, but relatively few develop seed heads. The seed head may be 15–20 cm long, with branches clinging to the central column of the stem, and is readily recognized because there are numerous breaks that destroy its symmetry. The seed is less than 6 mm long and is surrounded by numerous white hairs that extend most of its length.

Northern reed grass leafage is relatively palatable, but the stem is seldom eaten. Watching cattle and horses grazing over a marsh or wet meadow, one sees the stems standing upright, with the livestock grazing around the base of the plant. Chemical analyses indicate a protein content of 17–19% early in the season, but it drops to about 7% in August. The phosphorus content of northern reed grass is fairly high, about 0.2% in spring and early summer, and drops only slowly in late summer.

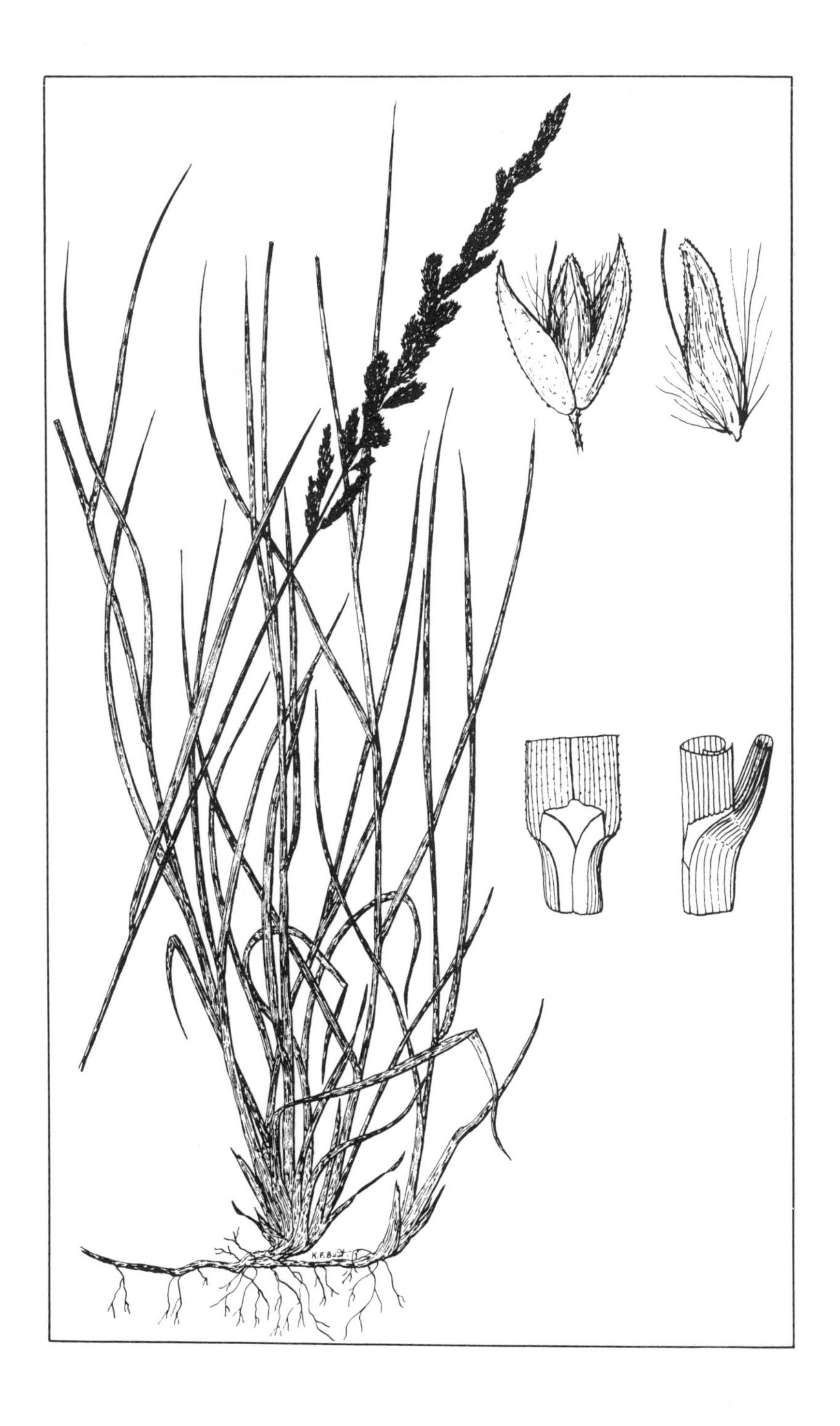

Pine grass *Calamagrostis rubescens* Buckl.

Pine grass, or pine reed grass, grows in fairly dense stands in open pine, aspen, and fir woods, and in sparser stands under dense pine forests. Its range includes the four western provinces in Canada, and southward to Colorado and California. It grows from sea level to the top of the timber line but prefers open forest sites at intermediate levels. It is common in the Cypress Hills, the Rocky Mountains, and central British Columbia, and occasionally in the northern boreal forests.

Pine grass grows vigorously from creeping rootstocks. Numerous tufts of leaves are often sufficiently dense to form a thick carpet. The root system is tough and very resistant to tramping, grazing, and water erosion. Although most new growth comes from the creeping rootstocks, viable seed is produced in thick heads on leafy stems 0.7-1.2 m tall. The reddish color of the seed head, particularly in open sites, helps to distinguish pine grass from its associated species. In general, the seed stems are relatively few in comparison with the basal leafage.

Pine grass is none too palatable except in the spring, when growth is lush and tender. Close herding forces stock to graze it, but associated nutritive and palatable plants such as vetches and asters are grazed out in such cases, without damaging the resistant pine grass. Thus, management practices have to be designed to protect associated species rather than the vigorous pine grass. Chemical analyses reveal a relatively low protein content at all seasons; even in the early leaf stage the protein content is less than 12%, whereas by October or November there may be no more than 2% protein in the plant. Fiber is high, and fat, phosphorus, and digestible carbohydrates are lower than in good range grasses. However, its wide distribution and rapid growth make it an important range forage in areas where pine trees dominate the vegetation.

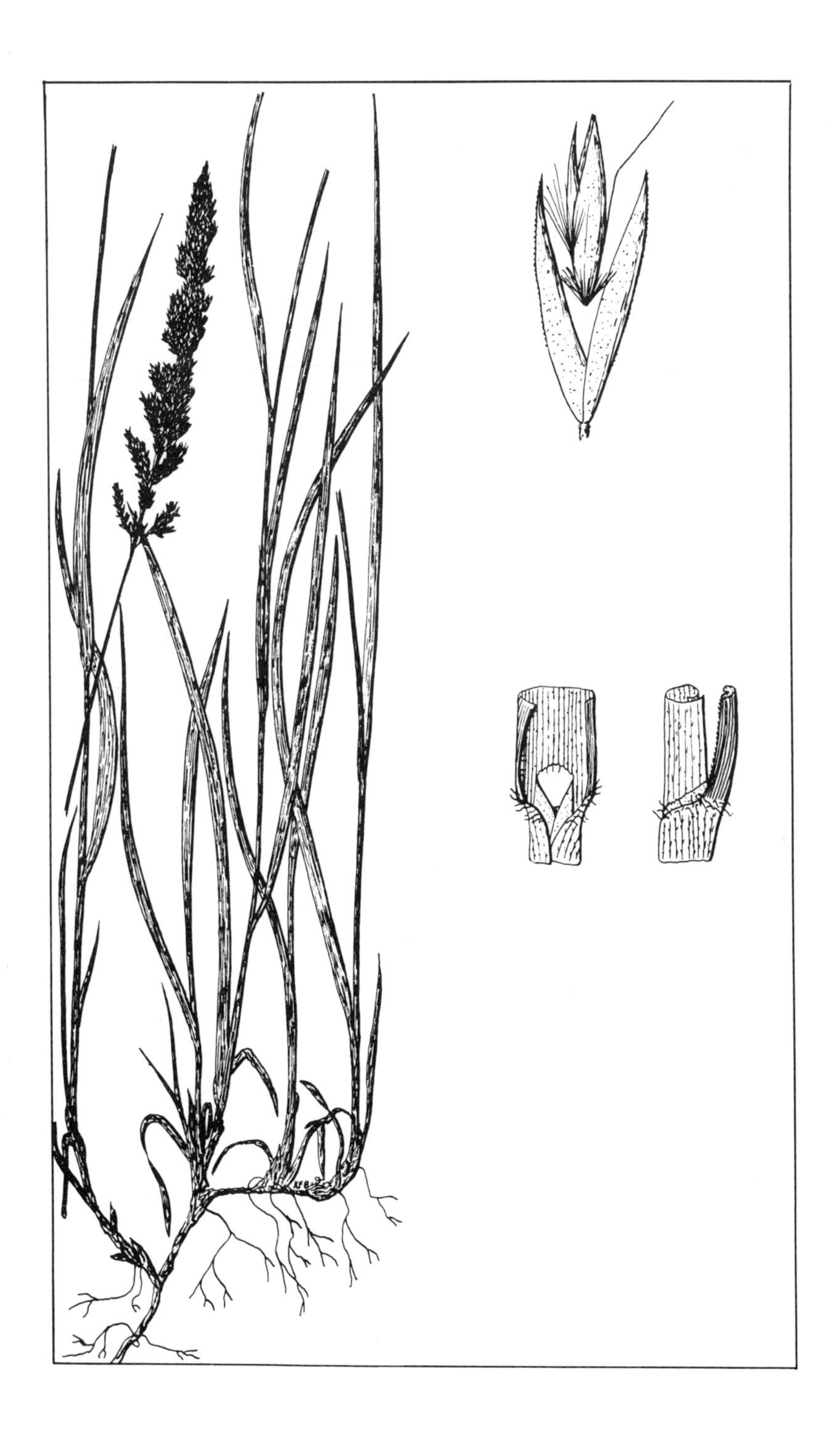

Plains reed grass *Calamagrostis montanensis* Scribn.

Plains reed grass has its center of distribution in southern Alberta, and extends sparsely to the center of Saskatchewan and Alberta. It is found also throughout Montana and as far south as Utah. Its habitat includes dry prairie and benchlands, and areas where western wheat grass, needle-and-thread, and silver sagebrush abound.

Plains reed grass has slender creeping rootstocks that send up single leaves or stems at intervals. There is very little basal leafage and seldom more than two leaves on the stem, both of which cling closely. Superficially and prior to flowering, plains reed grass resembles western wheat grass, but the leaves of western wheat grass, which grow at 45° to the stem, distinguish these two. The stem is erect, harsh, and seldom over 50 cm tall including the 7–10 cm flower head. During its flowering and seed-setting stages, plains reed grass resembles June grass but can be differentiated by the ring of white hairs surrounding the seed hull. Also, it usually grows as single stems rather than in tufts, and seldom forms stands of more than a few square metres.

Chemical analyses show a protein content of about 12% in early growth, rapidly dropping to less than 9% by midsummer. Its palatability decreases as soon as it reaches the boot stage.

K. F. Best

Sand dropseed *Sporobolus cryptandrus* (Torr.) A. Gray

The dropseed group contains nearly 40 species native to North America. However, they are generally southern grasses and only three are known to occur in the Canadian section of the northern Interior Plains. The group name *Sporobolus* means to cast or drop a seed and refers to the habit of shedding seed before maturity.

Sand dropseed has a wide distribution throughout southern Canada, the United States, and northern Mexico. In the prairie area it is often the dominant grass on stabilized sand dunes, and sparse stands are found in all areas with sandy soils. Heaviest-known stands occur in the Great Sand Hills and along the Qu'Appelle River. It is the most abundant and only important dropseed growing in the prairie area.

Sand dropseed is a small bunch grass with roots that extend to a depth of 1.25 m. Its leaves have a tuft of white hairs where the sheath and blade meet. It has solid stems that are often 1 m tall. The seed head is a closed panicle when it emerges from the leaf, but it opens into a pyramidal inflorescence at flowering. The panicle has single branches at each node; each spikelet contains one seed only.

Sand dropseed is rated as highly palatable in the southern states, but only moderately so in the prairie area. It starts growth about mid-May, later than associated species, and matures its seed before the end of July, after which the leaves cure. Unlike most grasses, sand dropseed is unpalatable after curing occurs. Protein content is not very high, only 10–11% in early summer, and drops to 5% in the fall. Its fiber content is higher than that of most other prairie grasses. An important character of the seed is its long life. Experiments show a high viability even after 20 years of storage.

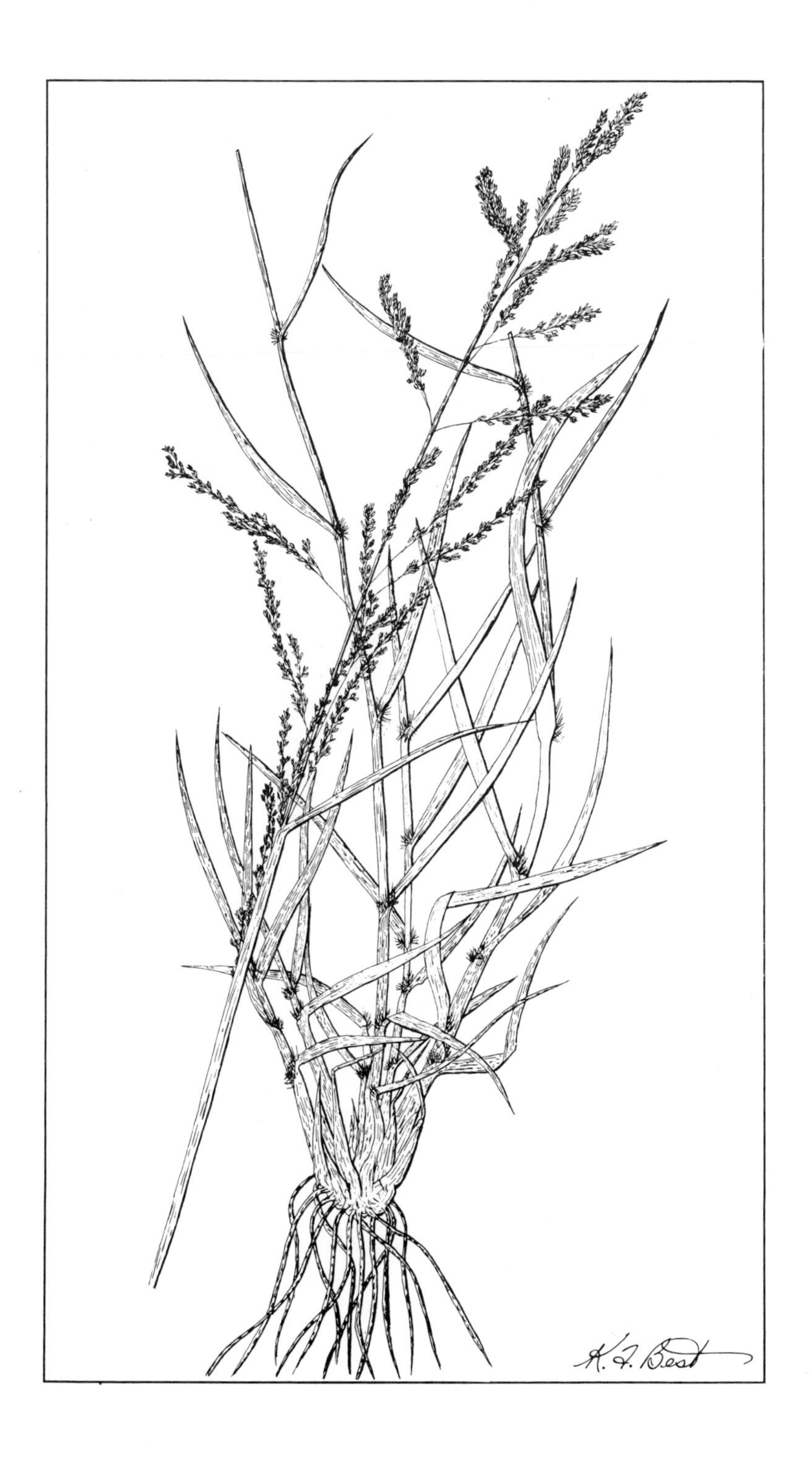
K. F. Best

Sand grass *Calamovilfa longifolia* (Hook.) Scribn.

Sand grass is the only Canadian representative of the genus *Calamovilfa.* It grows on sandy prairie and sand dunes, along lake shores, and in open forests. Its range spreads from the Rocky Mountains east to the Lake Superior region and southward to New Mexico and Kansas.

Sand grass has strong, scaly creeping rootstocks. These send down a mass of feeding roots to a depth of about 1.5 m, although most of them are found within 50 cm of the surface. The surface roots are dense and wiry and are well adapted to keeping the sandy soil from blowing. Considerable harsh basal leafage surrounds the coarse, leafy stems, which may be up to 2 m tall. Each stem ends in a feathery panicle, which may be 15-50 cm long. The seeds are enclosed in a hull, surrounded at the base with white hairs. Leaf sheaths of the previous year's growth surround the base of the stem and enlarge that section of the plant.

Sand grass is palatable during its first month of growth in the spring and after it cures on the stem in the autumn. At both periods leafage is readily eaten, and during the autumn livestock seek the plump and nutritious seeds. Stems are rarely eaten when other food is available. Despite its heavy root development, this species is susceptible to trampling and disappears from sites where livestock congregate. Because sand grass is a valuable soil-binding plant on land that erodes easily, care should be taken to maintain all existing stands. It is considered good winter pasture. Percentage protein content drops from about 16% during May to about 4% in November. Crude fiber content does not change much during the season, but the available carbohydrates increase from about 45% to 55% from May to October. Despite the sandy habitat of sand grass, it does not have a high ash content.

Slough grass *Beckmannia syzigachne* (Steud.) Fern.

Slough grass is one of two species of *Beckmannia*; the other species occurs in Eurasia.

It is a common grass in shallow sloughs and marshes. It prefers a clay soil covered with a shallow layer of organic matter and a sufficient depth of water to prevent the slough from drying before June 1. Its range includes all of the United States except the southeast corner, and all of southern Canada, Alaska, and a considerable portion of eastern Asia.

Slough grass is a stout annual. Its shallow roots support a leafy stem, which may be as tall as 1 m. A few basal leaves occur also, and one can see crossveins in them that cut the longitudinal veins at right angles to form rectangles. The branched head is classified as a closed panicle. Each branch carries one to four sickle-shaped spikes that produce 10–30 hard-coated seeds. The seeds can be stripped easily by drawing the head of the plant between the thumb and fingers.

Slough grass is palatable to cattle and horses from early stages of growth until the seed is nearly ripe. Chemical analyses indicate that nutrient elements are in good balance and that the seasonal protein trend is similar but slightly lower than that of upland grasses, dropping from about 14% in early June to 5% or less by mid-August. As with most grasses that grow in water, slough grass makes palatable and fairly nutritious but very light hay.

K. F. Best

Spangletop *Scolochloa festucacea* (Willd.) Link

Spangletop is the only species of *Scolochloa*. It is known also as water fescue. It grows along streams and shallow marshes from Manitoba to British Columbia in Canada and throughout Siberia and northern Europe. Heavy stands occur in Western Canada in the Qu'Appelle Valley, in Rush Lake near Swift Current, and in small sloughs and shallow lakes throughout the prairie area.

Erect, coarse, leafy stems grow from white, succulent creeping rootstocks. Some rough, long, green basal leafage develops. The straw-colored stems 1-1.5 m tall each bear a feathery, whitish panicle that may be 30 cm long. The oat-shaped seeds, 6-10 mm long, drop from the panicle before maturity. The whitish color of the empty heads identifies spangletop from a distance.

Spangletop requires a continuously moist soil for development and preferably a deep layer of organic matter on top of the subsoil. It should be flooded until mid-June. Where flooding lasts for a longer period, certain sedges, manna grasses, and cattail replace it. When flooding is for a much shorter period, either foxtail barley, desert salt grass, or western wheat grass dominates.

Spangletop makes quite palatable pasture and hay. It can be grazed fairly heavily, provided it is rested occasionally. However, because of its habitat, it is seldom available for early summer pasture. It should be cut for hay as soon as the water recedes and the land becomes dry enough to carry machinery. Even when cut for hay early in the season, it is very wasty because livestock do not readily eat the long, harsh stems. As with all water-loving grasses, its forage is soft and easily destroyed by frost, and it does not cure on the stem. Protein content reaches about 13% in early summer and drops to 7-8% in the fall.

K. F. Best

Sweet grass *Hierochloe odorata* (L.) Beauvois

Three species of *Hierochloe* grow in North America, but the one known as sweet grass is the only representative in the prairie area. It has a wide distribution between Labrador and Alaska and southward to Mexico; it is found also in Europe and Asia. The generic name *Hierochloe* is the Greek holy grass and refers to its use in religious festivals as a covering for floors and porches of churches. In the prairie area, sweet grass occurs naturally in low-lying upland, in draws, and throughout abandoned fields.

Sweet grass grows in open stands from slender creeping rootstocks, as well as from seed. Its feeding roots are shallow but numerous. Tufts of short, flat basal leaves surround the stem. The seed head is a shiny, open, pyramidal panicle. Each branch of the panicle produces several florets, which in turn enclose one perfect and two male flowers. The single seed in each floret drops to the ground at maturity, leaving the shiny, bronze glumes on the stem. The entire plant has a sweet odor.

Sweet grass is eaten fairly readily but is not as palatable as most range grasses. It adds a distinctive and pleasant odor to upland hay and does not transmit its odor to milk. Because its creeping rootstocks fill the soil, it is an excellent grass to help control water erosion.

K. F. Best

Switch grass *Panicum virgatum* L.

Switch grass is a member of the panic grasses, which belong to the millet tribe of plants. There are over 500 species, growing mainly in the semitropical regions of the world.

Switch grass is the only member of the panic grasses that is of importance in the prairie area, and then only in Manitoba and southeastern Saskatchewan. It is also common throughout Quebec and Ontario, as well as in the United States, Mexico, and Central America. It prefers low-lying, moist, light loam soils but does produce rank growth along streams in districts where it is common.

Switch grass is a perennial sod-former growing from short, stout, scaly creeping rootstocks. Because of its short rootstocks, the stems are crowded and thus a plant often looks more like a bunch grass than a creeper. It has few basal leaves, but long, somewhat bluish stem leaves occur from its base to its panicle. The panicle is feathery and nodding. The single seed of each spikelet is small and hard. Germination is usually poor. Strains have been selected for regrassing purposes in Kansas and New Mexico.

Switch grass is palatable when young and is readily grazed until the flowering stage. After this date its harsh, woody stems and leaves are not attractive to livestock when other grasses are available. Good hay can be secured by cutting this grass before the panicle emerges. This grass was of considerable economic importance during the period of settlement in Manitoba, but it is of little importance today because many of its extensive stands have disappeared into cultivated fields and overgrazing has reduced its abundance on native sod.

Tall manna grass *Glyceria grandis* S. Wats.

Tall manna grass is the most common of the six manna grasses occurring in the prairie area. It can be found throughout Canada and northern United States. It is a close relative of a Eurasian species and is considered a variety of that species by some botanists.

Tall manna grass is a plant of slough and lake margins as well as of marshy areas, where it grows in shallow water. It usually occurs together with several sedges and rushes.

Tall manna grass may grow to a height of 1.5 m from a shallow creeping root system. Its leaves are firm, growing to 30 cm long and often 15 mm wide. The numerous flowers are often purplish, borne in a feathery panicle that may be 60 cm long. The seeds are small and hard, with four to seven seeds in each spikelet. The seeds shell easily and drop from the plant at maturity. The seed hulls have very prominent nerves, a character that helps identify the manna grasses generally.

Tall manna grass and the closely related fowl and northern manna grasses, *G. striata* (Lam.) Hitchc. and *G. borealis* (Nash) Batch., are generally palatable and, because they grow in hard-bottomed, shallow sloughs, are available for pasture. The heavy seed crop assures the maintenance of stands unless tramping destroys the water-holding capacity of the soil or grazing is so heavy that seed crops cannot be matured. The seed and panicle are eaten readily, and careful pasturing is required to ensure seed replenishment. Its high protein content of 22% until mid-June drops rapidly to 7% by early August. It is low in fiber and high in phosphorus at all times. As with all marsh or slough grasses, its hay is light and wasty.

A.C. Budd

Alpine timothy *Phleum alpinum* L.

Alpine timothy is a worldwide species, growing in mountain bogs and meadows in North America, Europe, Asia, and Africa. Within the prairie area it has been collected only in the Rocky Mountains and the Cypress Hills. It has been suggested that the North American type is a different species from that found in Europe. However, the differences are small and most botanists prefer to consider them one species.

Alpine timothy is a shallow-rooted perennial with short creeping rootstocks. It has fairly dense basal leafage, which is soft and light green. The stems are up to 60 cm tall, recline at their base, and carry a few leaves. The panicle or seed head is tightly closed, about 3 cm long, somewhat egg shaped, and bristly. The seed is yellowish, enclosed in urn-shaped, awned husks, and slightly smaller than that of the cultivated species.

Alpine timothy is eaten readily by all classes of livestock. It makes a relatively early growth in view of its naturally cool habitat. It is entirely a pasture grass, not only because it grows in places where harvesting equipment would be useless, but also because its low growth habit makes it difficult to mow and gather. It has a somewhat higher protein content in later summer than its cultivated namesake.

K. F. Best

Timothy *Phleum pratense* L.

Timothy was imported into North America in mixed hay or in ballast of ships. However, it was adapted to the New England states and spread quickly. Its forage value was recognized as early as 1750, when seed was collected from wild stands and sown in cultivated fields. As with all new grasses it had many local common names, but it retained the name timothy from one of the early producers, Timothy Hansen. It moved from New England into Eastern Canada before 1800 and westward as the country was settled.

Timothy has roots extending to more than 1.25 m. Its crown consists of bulblike or swollen sections that each produce a mass of basal leafage and usually one leafy stem that ends in a seed head. All leaves are soft, light green, and 5-15 cm long. The seed head is a cylindrical panicle, which may be 15 cm long. The single seed of each spikelet is small; it is enclosed in an awned, urn-shaped husk.

Timothy is palatable as pasture and hay. During early growth, its protein content is as high as 20% but drops rapidly to 7-8% by the time seed is mature and to 4-5% by late fall. Of all grasses studied in an irrigated pasture test at Swift Current, timothy was the most palatable during its leaf and early stem stage.

Timothy requires considerable moisture supplies to maintain its stand. When annual precipitation is less than 400 mm, other grasses produce more and probably better pasture or hay than timothy does. However, in areas where moisture is sufficient, it spreads from cultivated fields into adjacent rangeland. This spread has occurred in Manitoba, along the northern edge of the prairie area, and in the Foothills. Unfortunately, many of these stands developed from seed much poorer than is available today; thus the quality of the feed produced is lower than that from strains selected recently. Most stockmen like timothy, particularly for spring and fall pasture.

K. F. Best

Tufted hair grass *Deschampsia caespitosa* (L.) Beauvois

Although six species of hair grass occur in North America, only tufted hair grass is widespread and abundant. It occurs in the Rocky and Appalachian mountains in the United States and northward into all provinces of Canada, as well as Iceland. In the drier central portion of the prairie area, it grows sparsely in marshes or around shallow sloughs, whereas at the boundaries of the area and in the Cypress Hills, tufted hair grass may be the most important species in all permanently moist sites such as marshes, springs, and narrow ravines. It seldom grows under trees.

Tufted hair grass is a bunch grass, depending entirely on seed to maintain its stands. It has dense, shallow roots, and a mass of deep green leaves covers the crown. The leaves are folded, each with a noticeable swelling where the sheath and blade join. One to twenty straw-colored stems 0.5-1 m tall are produced by each plant. The head is a feathery panicle that has several branches growing in whorls of six or ten at points 2-3 cm apart. There are two seeds in each spikelet.

Tufted hair grass is rated as a highly palatable species that is resistant to grazing. It starts growth early in the spring and its leaves remain green throughout the summer. Very lush growth, as well as the mature leaves that are coarse and wiry, may be refused by livestock. Management practices should allow seed to set in abundance because the plants are short lived and reproduce only from seed. Chemical analyses show a high protein content of up to 18% in May, decreasing to 7% by August. Fiber is relatively low and phosphorus content is well above average. Tufted hair grass is a valuable range grass that should be preserved wherever it grows.

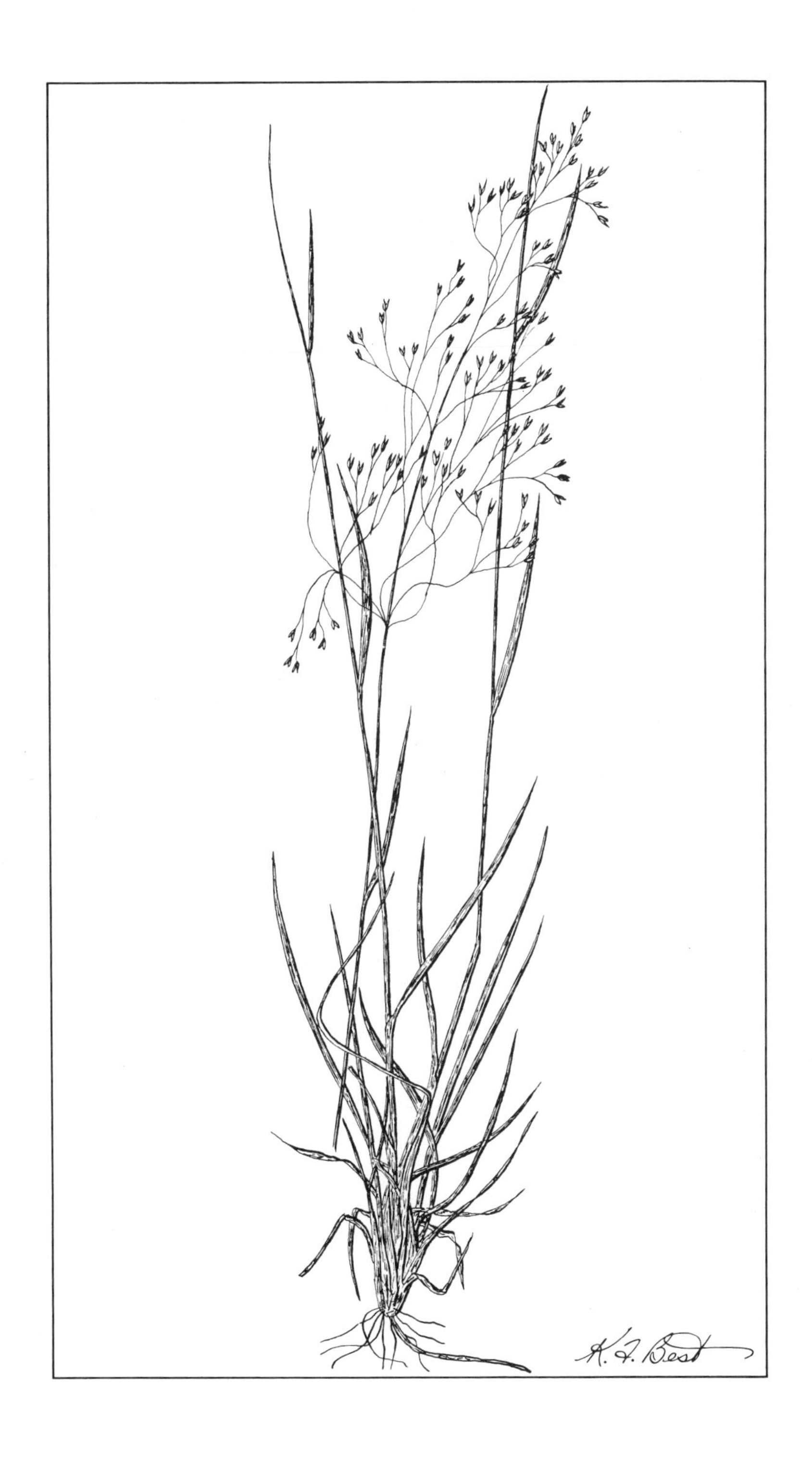

Awned wheat grass *Agropyron trachycaulum* var. *unilaterale* (Cassidy) Malte

Awned wheat grass occurs throughout the range area. It is similar to slender wheat grass in growth characters, except its awns are longer and its seeds are clustered on one side of the spike. It prefers moist, well-drained, nonalkaline loam soils and is more common in the western and northern portions of the area than in the dry central zone.

The basal leafage of the species is soft and very palatable to all livestock. However, the stem and seed heads are rarely eaten. Thus, single stalks, or two or three together, are not uncommon, growing strongly in heavily grazed fields. The unpalatability of the stalk and head ensures a supply of seed to establish new plants. With no creeping rootstocks this species propagates itself by seed alone, so the unpalatable seed character is a fortunate one for range maintenance. This grass does not produce a desirable hay because of the unpalatability of the straw and spikes. The nutrient content of awned wheat grass follows the same pattern as that of slender wheat grass.

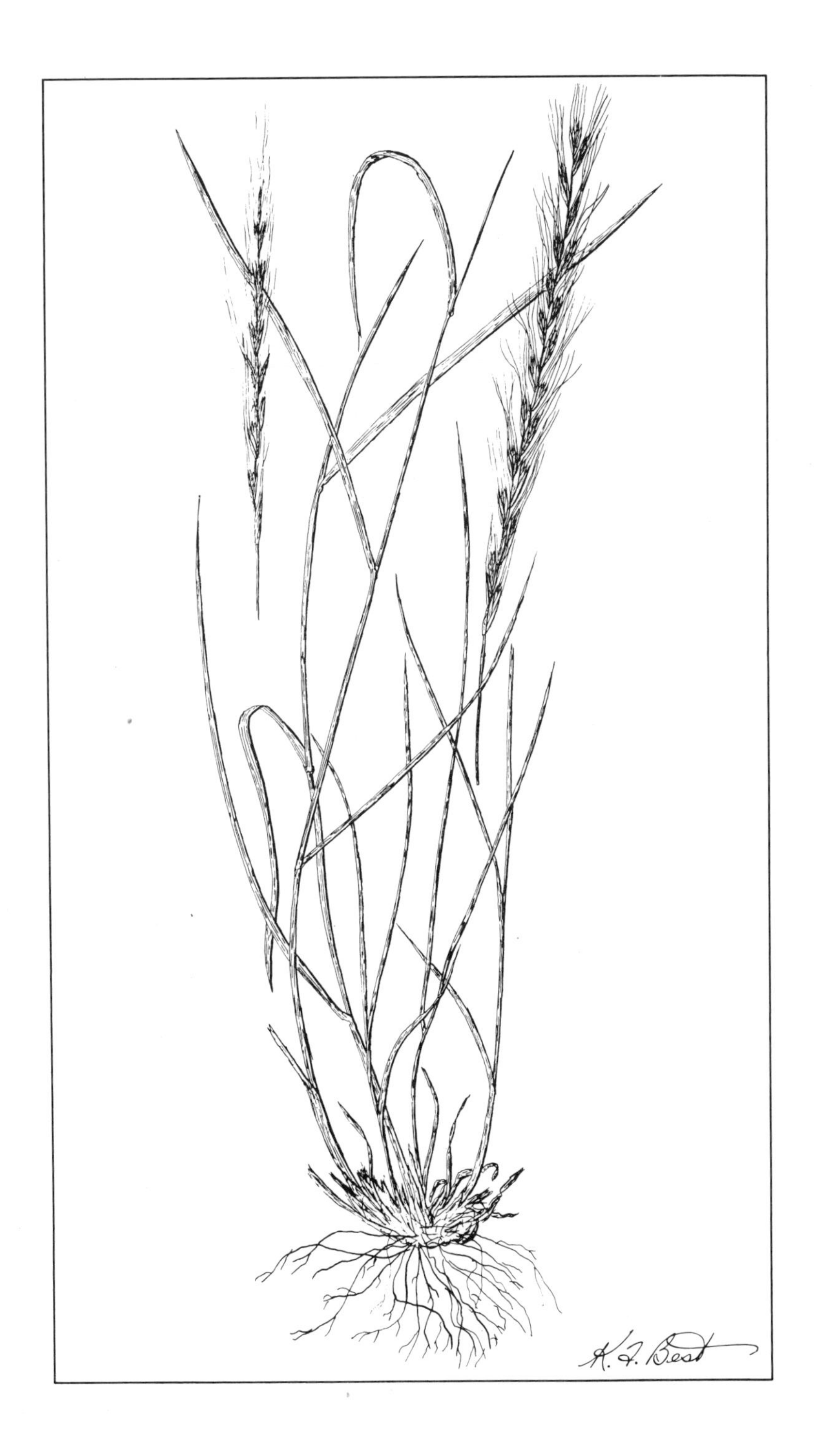
K. F. Best

Crested wheat grass *Agropyron cristatum* (L.) Gaertn.

Crested wheat grass was introduced into Canada in 1915 from western Siberia. It was well adapted and has been used for pasture and hay production and for regrassing abandoned and eroded land in the drier sections of the region. Although it grows well on all soils, its excellent response on sandy land indicates its particular suitability for that soil type. In the Foothills and Cypress Hills, as well as adjacent to the northern boundary of the area, it is less productive than the brome grasses, timothy, or the fescues.

Crested wheat grass is a bunch grass, with dense roots that grow to a depth of 30 cm. Its leaf growth is rapid and steady from mid-April to late June but ends by mid-July when the seed starts to ripen. Fall growth may occur during September. Highest yields are obtained from 3- or 4-year-old stands, after which yields are usually lower as stands become progressively sod-bound.

Because of its rapid growth during April and May, crested wheat grass, either alone or with alfalfa, makes excellent spring pasture. Further, it can be grazed heavily until mid-June without injuring the stand or reducing its long-time yielding capacity. Lighter rates of grazing can be practised throughout the summer. Seeded with alfalfa, in 30-cm rows, crested wheat grass makes excellent hay when cut before the end of June. It recovers rapidly from drought.

Crested wheat grass is palatable to all livestock, particularly so during May and June. In May and June its protein content drops from 23% to 12–14%, with further decreases to about 4.5% by October. Besides its satisfactory protein content during the spring, the crop has a good nutritive balance among protein, carbohydrates, vitamins, and minerals.

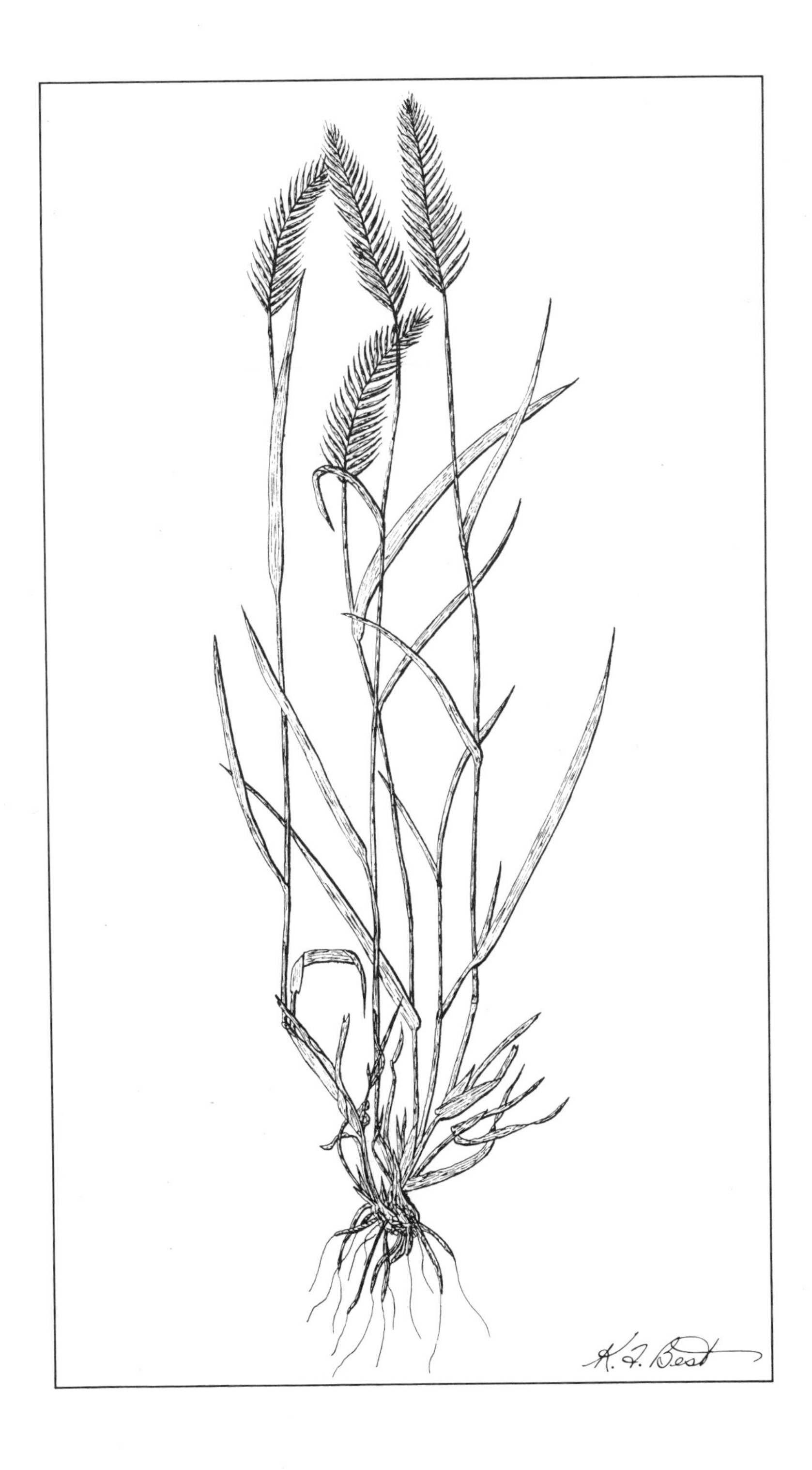
K. F. Best

Intermediate wheat grass

Agropyron intermedium (Host) Beauvois

Intermediate wheat grass is a native of eastern Europe. It was introduced into the United States in 1932 and into Canada a few years later. It is well adapted to a great variety of conditions in the western states, where rainfall is greater than 380 mm annually. In the prairie area it has demonstrated its ability to maintain stands where smooth brome or creeping red fescue do well. It is currently being studied by plant breeders who are trying to select high-producing as well as drought-tolerant and winterhardy strains.

Intermediate wheat grass has a deep feeding root system, as well as extensive, creeping rootstocks that form a tough sod. Considerable long, dark green basal leafage surrounds leafy stems 1-1.5 m long, ending in a long spike. The edges of many of the leaves have short hairs that help to distinguish this grass from others closely related. The seed head consists of spaced spikelets, one at each node, each containing two to six seeds nearly as large as oats. The glumes are about one-half the length of the spikelet and are pointed.

As intermediate wheat grass is new to many farmers and ranchers, few realize its potential value. It makes very palatable pasture in all seasons of the year. It has some curing properties and usually makes a fall growth that is sought by livestock. By itself, or in alfalfa mixtures, it produces good yields of high-quality, palatable hay. Seed crops are not heavy, 200-300 kg/ha; but the tall stems and large, firm seeds make harvesting with combines relatively easy. Seed is available from most seed houses. Its protein content is about 19% during early growth and drops to 15% at maturity.

Intermediate wheat grass can probably be useful for pasture and hay in all districts throughout the prairie area where annual precipitation exceeds 400 mm, or where the precipitation-to-evaporation ratio is about one. At present it is not recommended for the dry central area, except for hay under irrigation.

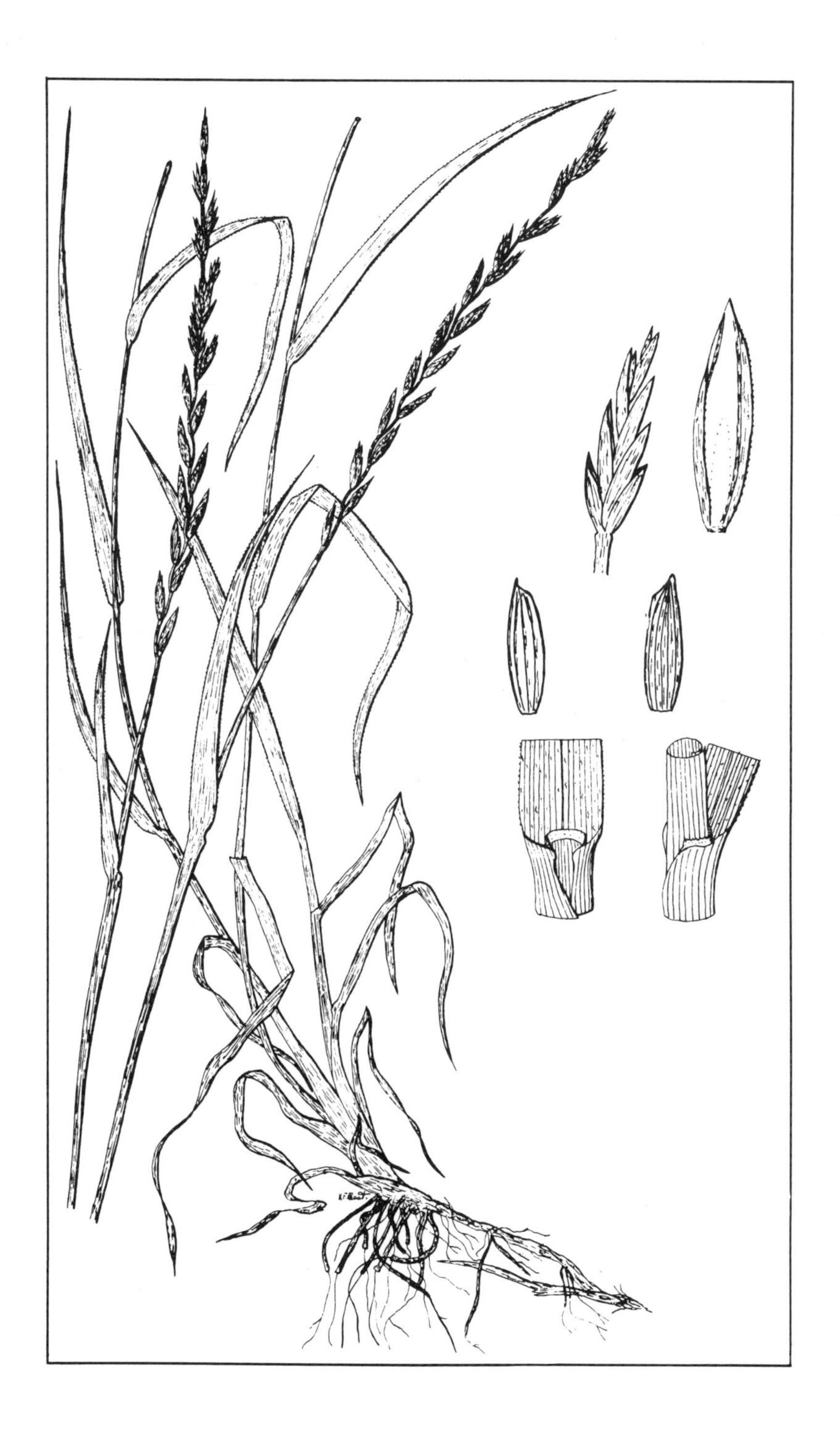

Northern wheat grass

Agropyron dasystachyum (Hook.) Scribn.

Northern wheat grass is similar to western wheat grass in growth habits and general appearance, but it can be distinguished by the lighter green color and tufted appearance of the basal leaves, and by the seed husk, which is covered with fine hairs. It grows in mixed and sparse stands with western wheat grass or needle-and-thread on clay and loam soils, and occasionally in nearly pure stands on sandy land. It grows throughout the entire region and is considered a valuable forage plant. It is known also as thickspike, Yukon, and downy wheat grass.

Northern wheat grass is equipped with a three-way root system: creeping, underground stems spread the plant and are sources of new stands; a very dense, shallow root system that penetrates to a depth of about 25 cm takes advantage of surface moisture; and a few deeper feeding roots may penetrate to a depth of more than 50 cm. This combination of root types, which feed and extend the stand, increases the tolerance of northern wheat grass to drought and its resistance to invading weeds. Dense stands of northern wheat grass seldom set seed, but individual plants do so during years when growth is able to continue for the 115 days required for the plant to reach maturity from emergence in spring.

The plant is palatable despite its fairly coarse stem and harsh leaves. Chemical analyses indicate that its nutrient content follows the same pattern as other dryland grasses. Protein decreases from about 16% in early May to about 4% in October, but the energy-producing elements such as fats, starches, and cellulose maintain a high level of about 45% from emergence to maturity. The species cures on the stem to produce palatable and nutritious winter pasture and hay.

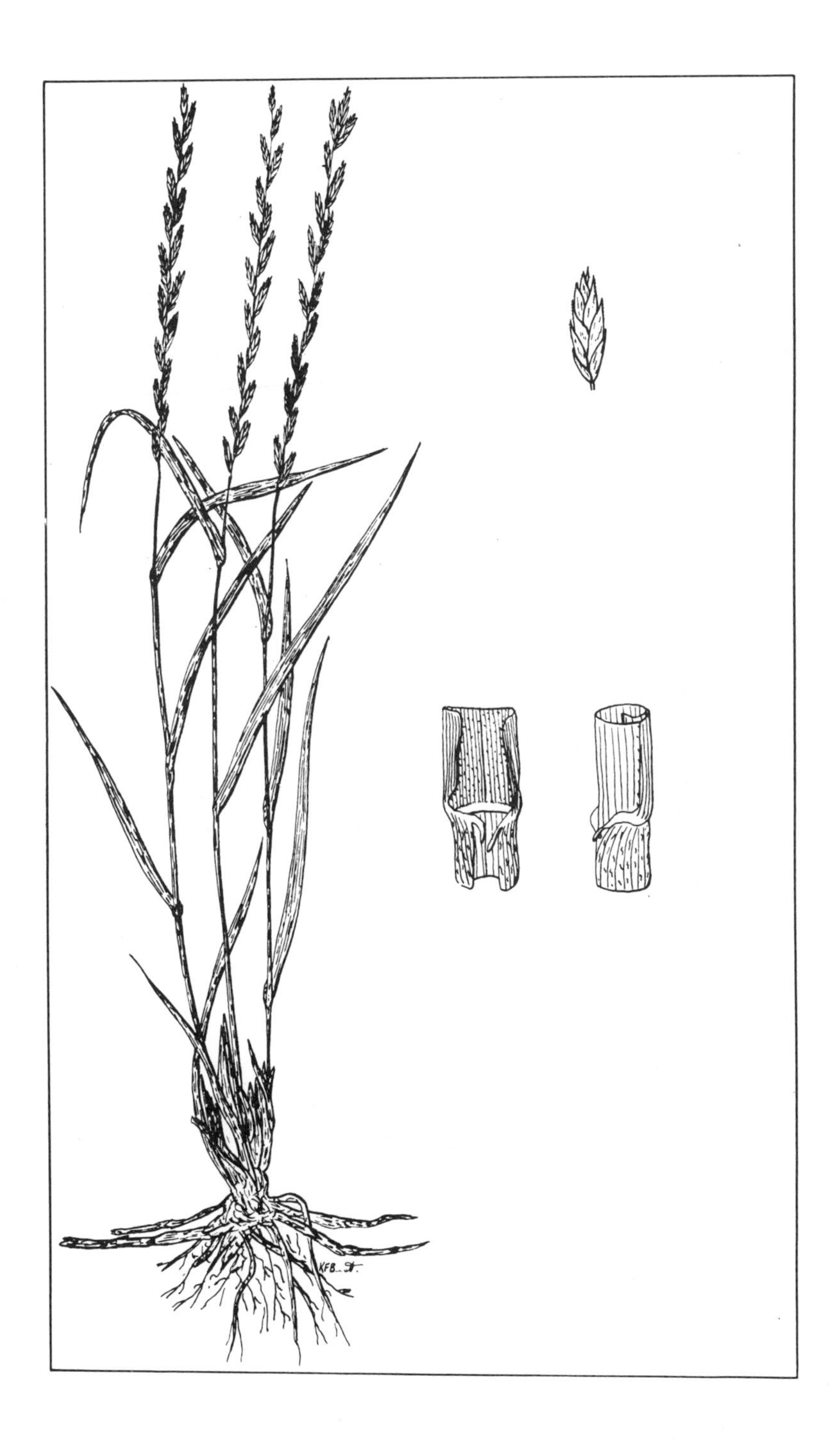
KFB. Jr.

Quack grass *Agropyron repens* (L.) Beauvois

Quack, or couch, grass is an introduced wheat grass from Eurasia. It has established itself in the northern and eastern portions of the prairie area in hay fields, abandoned farms, and crop land. It requires considerable moisture to persist and thus has not invaded the dry central core of the region except for low-lying areas.

As with the closely related western wheat grass, quack grass has a three-way root system. Scaly, yellow, creeping underground stems spread rapidly, and surface feeding roots fill the upper 50 cm with a dense mass of fiber; a few deeper feeding roots grow to a depth of 2.5–3 m. The leaves are green to yellowish green, ascending close to the stem, and often spiraled. Compact heads with overlapping seeds produce heavy seed crops on all young or sparse stands. A sod-bound condition develops in 3–5 years after establishment, and production declines accordingly.

This grass has a two-faced reputation. It is a noxious weed in cultivated fields but a valuable hay and pasture grass wherever it has established itself in rangeland. Farmers generally regard it as one of the most difficult weeds to eradicate from wheat fields, whereas stockmen recognize it as palatable, digestible, and nutritious pasture during the spring and a strong hay for winter feeding. Besides its forage quality, quack grass is a valuable soil-binding plant for stabilizing eroding land. Like other wheat grasses, quack grass cures on the stem. The few chemical analyses available indicate a similar protein trend to other wheat grasses but lower contents of energy foods and total nutrients. Two of its most useful characters for range purposes are its early emergence and rapid spring growth.

K. T. Best

Slender wheat grass *Agropyron trachycaulum* (Link) Malte var. *trachycaulum*

Slender wheat grass, also known as western rye, is a native of southern Canada and northern United States. It makes excellent growth on moist, well-drained soils and has some tolerance to alkali soils and drought. Unlike most other wheat grasses, its plants have a short life, seldom over 5 years.

Slender wheat grass has a leafy, bunch habit of growth, with dense, fibrous roots extending to a depth of 50 cm. It has no creeping roots but 2- or 3-year-old plants may increase their basal area by tillering. The seed stalks are erect, with spikes that are either dense or very open. A relatively consistent character is the reddish or purple color of the stems near the base. Short awns extend from the seed hulls.

Slender wheat grass yields well for its first 3 or 4 years of growth. It is fairly palatable and nutritious in all stages of maturity, and its seed is sought by livestock in the autumn. Chemical analyses indicate that it has a well-balanced nutrient composition. Heavy grazing reduces stands quickly, allowing them to be invaded by weeds. Slender wheat grass contains about 23% protein in early growth stages. This level drops to 15–16% by midsummer and to 8–10% by late summer and fall.

Selections of slender wheat grass have been grown as cultivated wheatgrasses throughout Canada. Because it develops its stands quickly, it has been used in rotation with annual crops and in mixtures with slower-developing but longer-lived grasses. It grows well with alfalfa, but soon disappears from a stand that contains crested wheat grass or quack grass. As a cultivated crop it is better adapted to the northern and eastern sections of the area than the dry central portion.

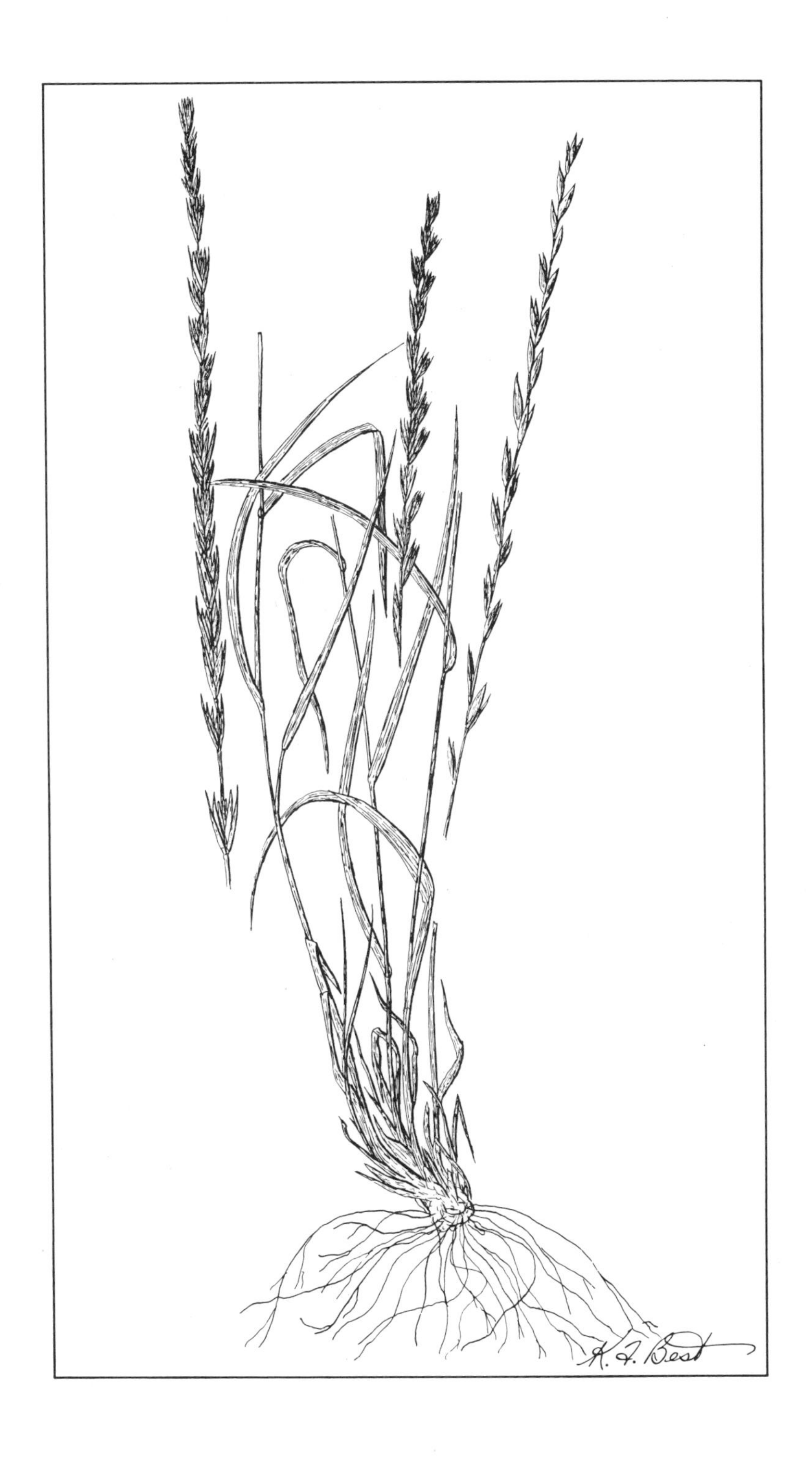

Streambank wheat grass *Agropyron riparium* Scribn. & J.G.

Streambank wheat grass is a native species, much resembling northern wheat grass, but with a more vigorous creeping root system, smooth instead of hairy flowers that are usually closely overlapping, and narrower leaves.

Known as streambank wheatgrass when cultivated, it is a special-purpose grass, well suited for use in erosion control in ditch and canal banks, waterways, roadsides, landing fields, and similar areas. Although it tolerates moist conditions well, it is not suited for irrigated areas.

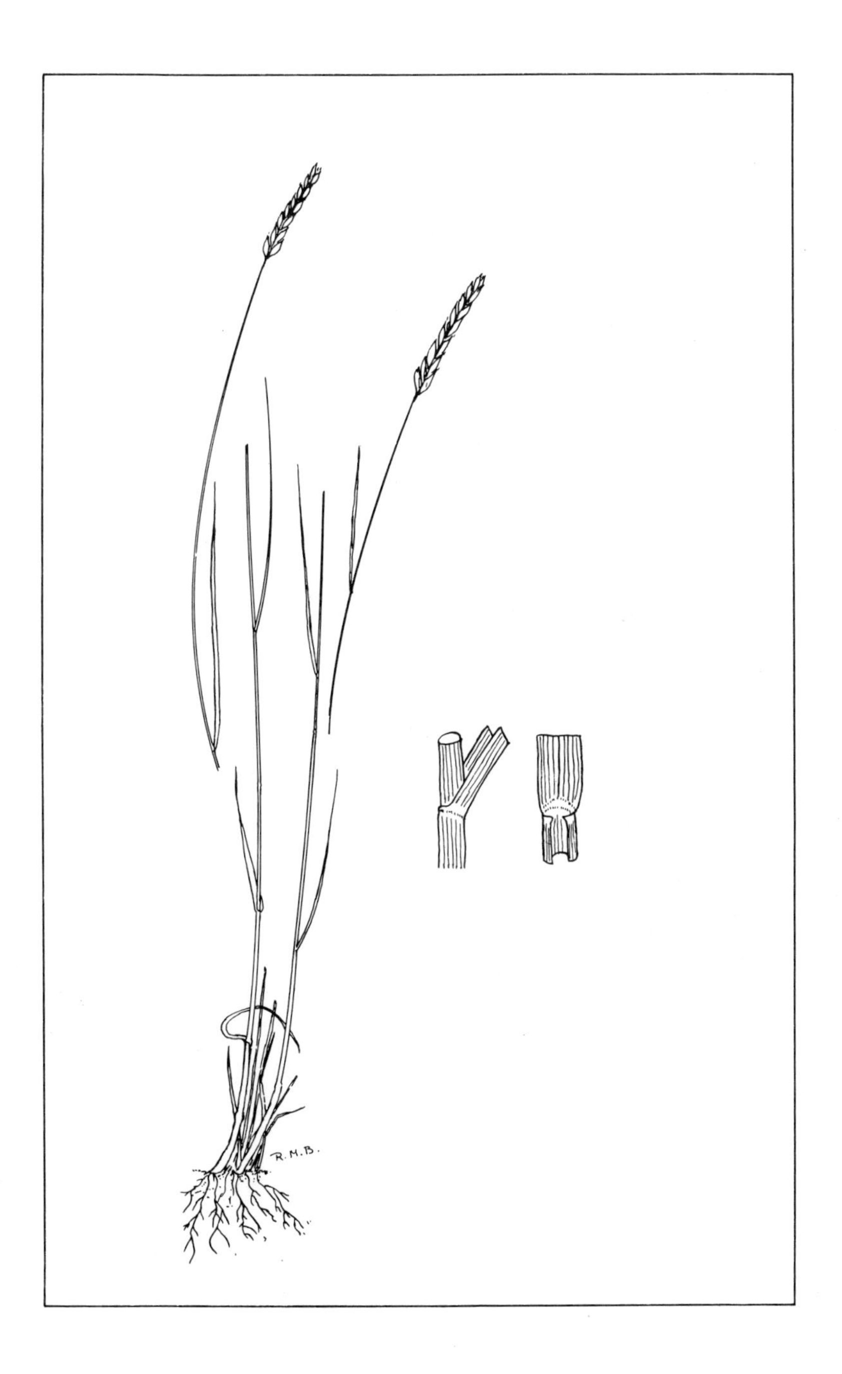
R.M.B.

Tall wheat grass *Agropyron elongatum* (Host) Beauvois

Tall wheat grass is another of the grasses introduced from southern Russia, known when cultivated as tall wheatgrass. It has been cultivated in the prairie area since 1929, when it was introduced into Canada by the University of Saskatchewan. For many years it created little interest because it is coarse, not particularly drought tolerant, and slow to establish. However, it has recently demonstrated its ability to thrive in subirrigated, saline soil where foxtail barley is usually the dominant grass.

Tall wheat grass is a bunch grass that extends its size by producing tufts on short rootstocks at the edge of the mature plant. Long, coarse, light green basal leaves surround several 1-2 m leafy stems. The spike is interrupted, in that the spikelets do not overlap as in western wheat grass. Each spikelet has a character that distinguishes tall wheat grass from other wheat grasses; the spikelets and the four to ten enclosed seeds grow away from the stem like a sickle. The seeds are large and heavy. The outer seed covering, the glume, is square across its top.

Tall wheat grass is more palatable than the coarse leaves and stems suggest. It is reported to be good pasture and to make excellent hay when cut shortly after heading. It yields well by itself or in mixture with alsike clover, despite the open spaces between the plants. Seed is harvested easily with a combine. Seeding in 30-cm rows at a rate of 12 kg/ha is recommended.

Tall wheat grass is a slow grower when young. However, it fills in its stand by establishment of new plants from seed and by making extensions from mature plants. One other important character is its ability to establish and maintain its stand on moist, moderately saline soils. It is more salt tolerant than either smooth brome or slender wheat grass and nearly as much so as foxtail barley. It does not develop in dry saline sites, nor does it grow where the soil is so saline that native weeds cannot persist. It is also useful in grass barriers for snow management.

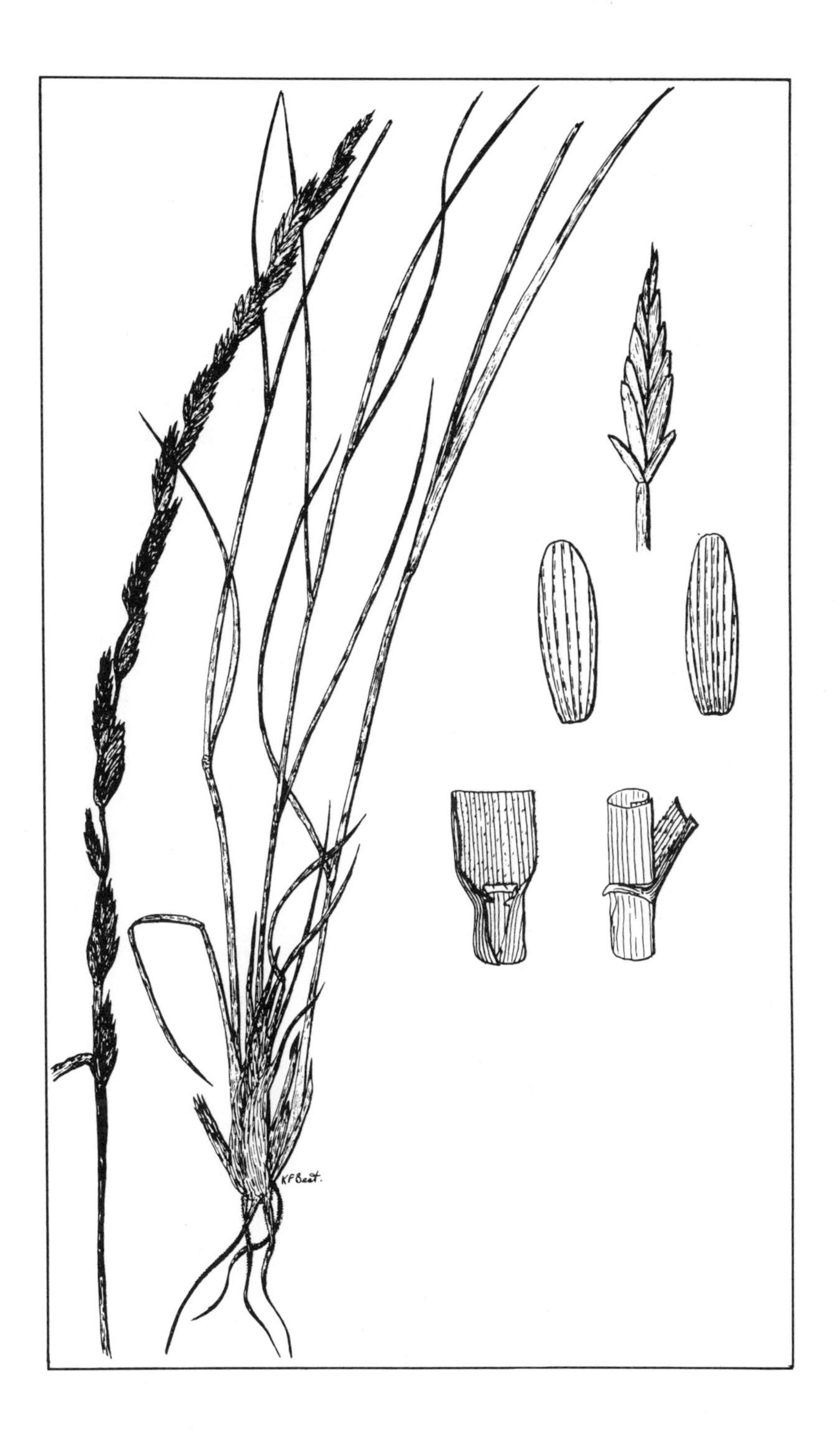
KF Best.

Western wheat grass *Agropyron smithii* Rydb.

Western wheat grass is known also as bluejoint, Colorado bluestem, western couch grass, and bluestem. It grows in fairly dense stands on clay soils in association with green needle grass, on saline soils with salt-tolerant grasses and greasewood, and in sparse stands among blue grama and needle-and-thread on upland sites. It is found throughout Western Canada from the Portage Plains to the west coast.

Western wheat grass leaves, which are rigid and blue green, grow at a 45° angle to the stem. The plant has creeping roots that assist in spreading its stands. Also, it has a well-developed root system to feed itself; there are not only deep feeding roots that penetrate to a depth of nearly 1.5 m but also a mass of surface roots feeding to a depth of 20–25 cm. Thus the plant is well adapted for surviving drought and for taking advantage of light rains that only soak the surface. Undoubtedly its three-way root system helps western wheat grass survive and produce during dry periods and to recover rapidly after prolonged droughts.

Western wheat grass is a fair yielder, producing about four times the yield of blue grama on upland and as much as 4000 kg/ha on sites flooded in spring. It rarely matures seed and requires about 110 days of growth for seed to ripen after the leaves emerge in spring. Despite its rather harsh leaves, it is fairly palatable, nutritious, and digestible. Like many other range grasses, it cures on the stem. Chemical analyses indicate a protein content of 18% in early May, which declines to about 3–4% by October; digestible carbohydrates increase their content from about 40% to 50% during the same period. Western wheat grass is the standard of quality for range hay and few, if any, native grasses equal it for this purpose. Feeding tests have demonstrated that well-cured hay cut in the late leaf stage has a 60% digestibility.

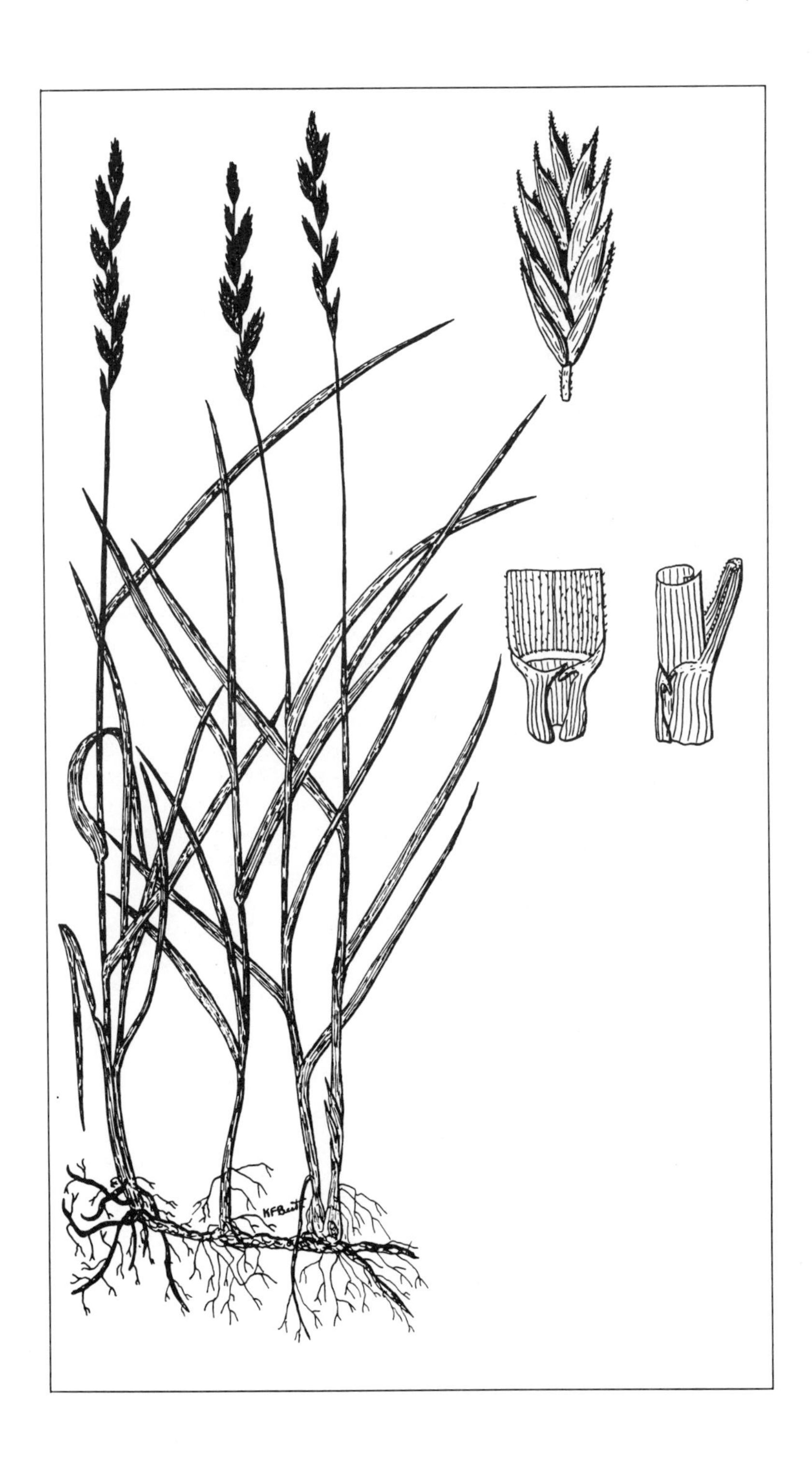

Altai wild rye grass *Elymus angustus* Trin.

Ten species of wild rye grass are native in the prairie area. None of these species is very palatable, but several are useful as soil binders.

Altai wild rye grass is a native of southern Siberia, where it is found in the steppes, semideserts, and saline areas from the Caucasus to the Altai Mountains. It is well adapted to the climatic and soil conditions of the prairie area and grows well on loam and clay soils. Its root system can penetrate to a depth of 3.5 m or even deeper. Its tolerance for salinity is equal to that of tall wheat grass.

Altai wild rye grass forms large clumps with many coarse, erect basal leaves that are light green to blue green. The leaves can reach a length of 40 cm and may be as wide as 12 mm. Stems are 60–120 cm tall, and the seed head is 15–20 cm long. Seeds are 10–15 mm long.

A cultivated variety, sold as Altai wild ryegrass, is valuable for fall and winter pasture. It retains its nutritive value better than other grasses and recovers well after grazing when moisture is available. Protein content ranges from 24% in the early leaf stages, through 18% in the shot blade stage, to 8% at maturity. Altai wild rye grass is not recommended for hay.

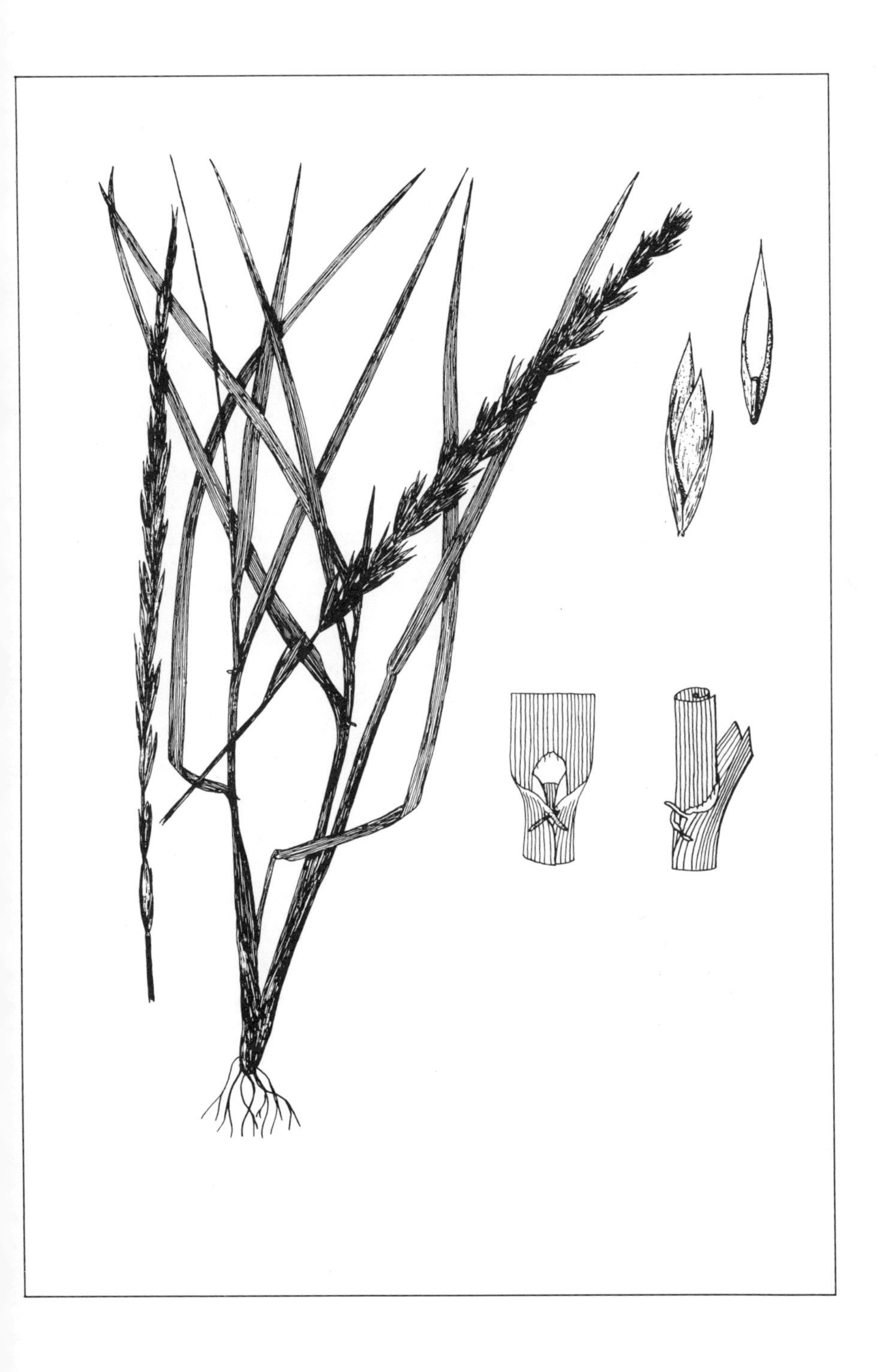

Canada wild rye *Elymus canadensis* L.

Canada wild rye, also known as nodding wild rye grass, is a native species. Its range includes all of North America. In the prairie area it is found in sparse stands on sandy soil and along river banks, roadsides, and other disturbed ground. Attempts have been made by selection and breeding to develop strains of this grass for improved pasture and hay. The best of these is known as Mandan wild ryegrass.

Canada wild rye grass has a bunch or tussock type of growth. Deep, wide-spreading roots anchor the plant firmly and provide nutrients for the yellowish green leaves and stem. There is considerable basal leafage, and the tall stems are leafy from their base to the spike. The spike itself is about 15 cm long and nodding. The spikelets, two at each node, are rough and long awned. The plant is a heavy seed producer, but special seed treatment is necessary to remove awns before the seed can be sown through a drill.

Canada wild rye is not very palatable, although it is eaten in the spring. Its mature growth has even been described as being poor for bedding, although hay of fair quality is obtained when harvested in the boot stage. Although livestock do eat this hay, they do not take to it readily and their gains are seldom satisfactory. Because the plants start growth late in spring, later than most natives, very little use is made of this species, even in its most palatable state. However, Canada wild rye establishes itself quickly and is a useful plant for controlling erosion. It competes successfully with crested wheat grass.

K. F. Best

Hairy wild rye grass *Elymus innovatus* Beal

Hairy wild rye grass grows from Alaska to the Red River in Manitoba, and southward to include the states of Montana, North and South Dakota, Idaho, and Washington. It develops dense stands under open poplar and pine sites where the soil is sandy or gravelly. In the prairie area it occurs within the parkbelt and northward, as well as in the Cypress Hills and throughout the lower ranges of the Rocky Mountains. It comes in quickly on all sites cleared of poplar and pine.

Hairy wild rye grass grows from slender creeping roots, as well as from the abundant seed supply. A deep and spreading root system supports the plant. There is considerable basal leafage, which is light green and hairy, but the tall straw-colored stems are relatively free from leaf growth. The hairy spikes are dense and nodding, growing as long as 15 cm. There are two spikelets at each node of the spike. These spikelets may produce two to six flowers, each of which may develop into a seed. The entire spike is very hairy, which gives a characteristic bluish or grayish color to the head.

Hairy wild rye grass is not palatable and is never eaten when other grasses are available. The coarse leaves and stems are not succulent at any time, and the hairiness of the entire plant undoubtedly discourages consumption. Nevertheless, the plant is useful because it comes in quickly on denuded sites and thus reduces wind and water erosion.

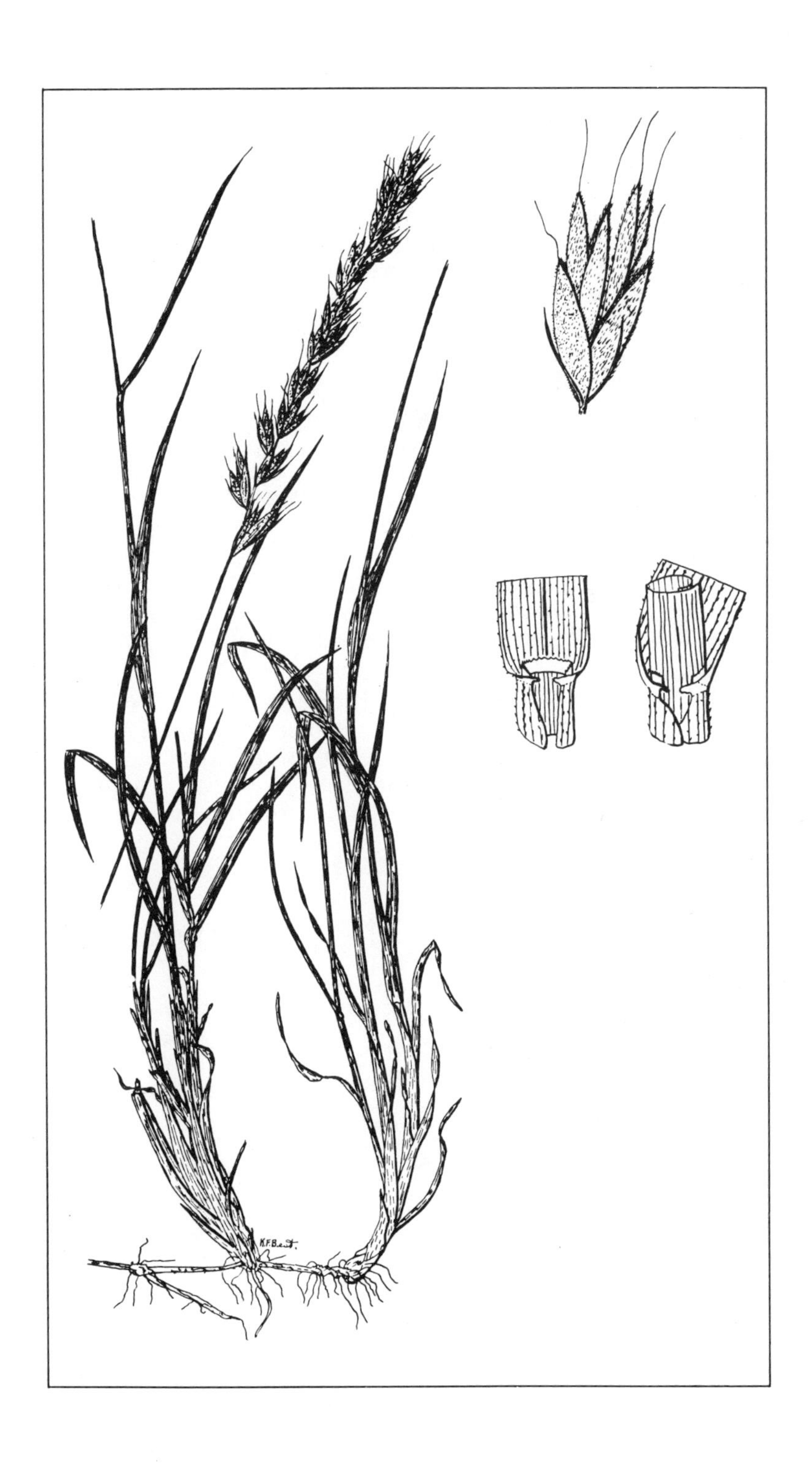

Russian wild rye grass *Elymus junceus* Fisch.

Russian wild rye grass, usually listed as Russian wild ryegrass, was introduced for cultivation into the United States some 40 years ago from its home in western Siberia. Within a few years it was being tested at experimental stations and universities in Western Canada. Because it is not a particularly high-yielding species for either hay or seed, its better qualities of drought and cold tolerance, high protein content, and autumn palatability were overlooked when grasses were being selected for pasture in the prairie area of Canada. Now, however, thousands of hectares in the Prairie Provinces have been cultivated to this species.

Russian wild rye grass is a deep-rooted bunch grass. It produces a large number of deep green basal leaves and a few leafless stems 0.6–1.2 m tall, which are straw colored when mature. The seed head is a spike with overlapping spikelets. There are two spikelets at each node, and each spikelet may contain one to four or more seeds.

Pastures of Russian wild rye grass are valuable for grazing in all seasons, but particularly in May and June, and again in mid-August to October. Its protein content in spring is very high, 21% or more, and remains high compared with other species in fall when it still may be more than 7%.

Russian wild rye grass is a pasture grass and is not recommended for hay. Seeding rates for pasture are 3.5–6.0 kg/ha at a depth of 1.5–2.5 cm. Row spacing in dry areas should be 45–90 cm, and in moist areas, 30–45 cm. For seed production, Russian wild rye grass should be seeded at 2.5–3.5 kg/ha in rows spaced at 90 cm. The aftermath provides excellent fall grazing.

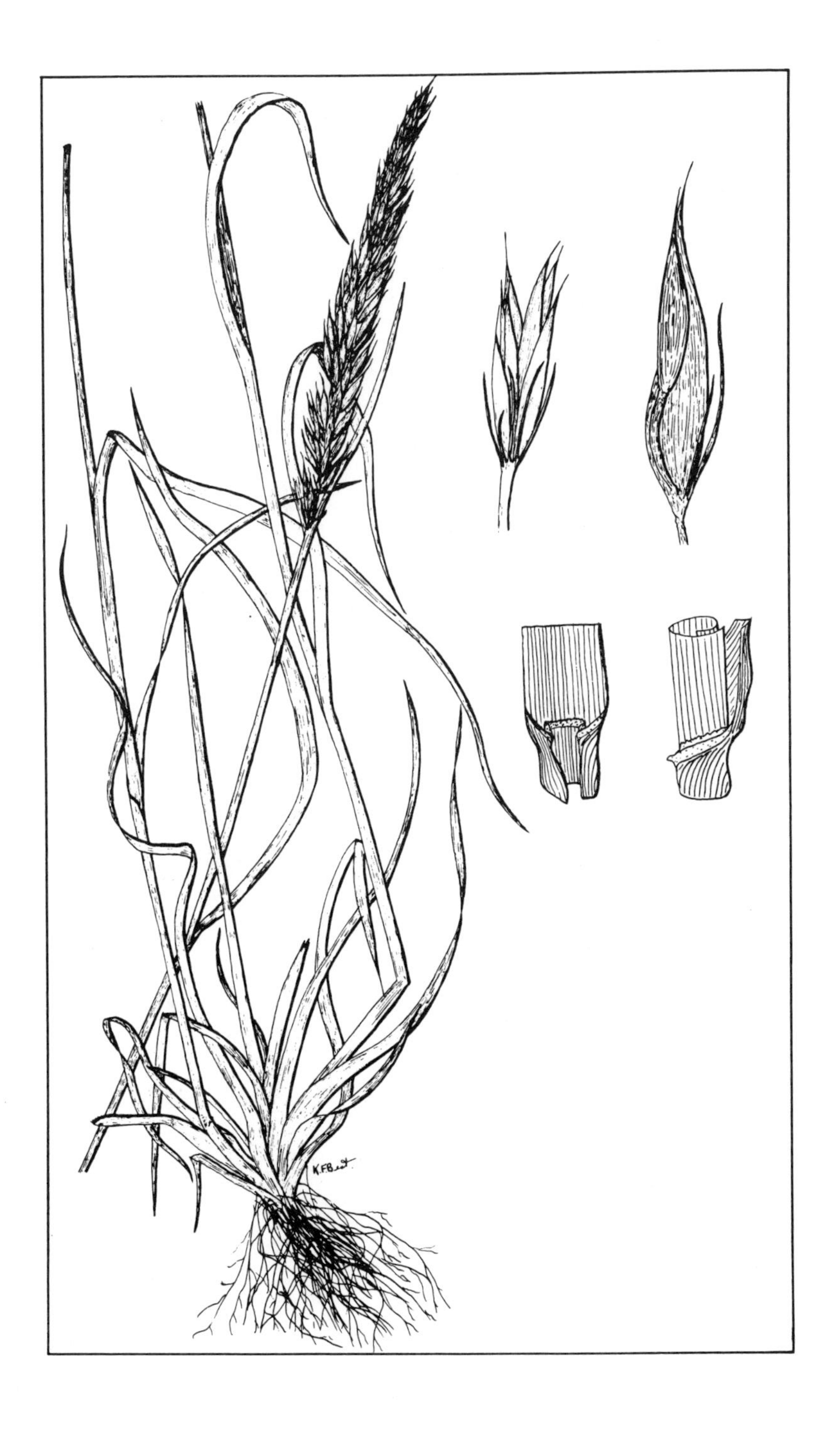

Virginia wild rye grass *Elymus virginicus* L.

Virginia wild rye grass has a wide distribution throughout North America. In the prairie area it is found along streams, under poplars, and throughout moist areas generally, although it is not resistant to ponding or excessive flooding.

Virginia wild rye grass is a bunch grass with a few basal leaves and a large number of stems, which carry long, limp leaves from the base to the spike. The plant has a shallow but dense root system, which fills up the soil and helps to establish stands quickly. The light green stems are often 1 m tall and are smooth and shiny. The spike is erect and bristly, often partly enclosed in the upper leaf. There are two spikelets at each node, and each spikelet contains two to five seeds. Awns 12-20 mm long extend from the tip of the shiny seed hulls.

Virginia wild rye grass is extremely variable in its appearance. Some forms have nearly awnless seed hulls, whereas others have very hairy seeds; the leaves may be either short and narrow or long and fairly wide. However, no matter the form, this grass has a low palatability rating. As with most other wild rye grasses, it has low protein and high fiber contents at all stages of growth, and a particularly low protein rating after autumn drying. It does not cure on the stem and is relatively short lived.

Plant studies have been undertaken to select desirable strains of Virginia wild rye grass and to improve these strains in plant breeding programs. This work has increased yields, palatability, and stand establishment. However, in areas where it grows, other grasses such as the brome grasses and timothy outyield it and produce more palatable and nutritious feed.

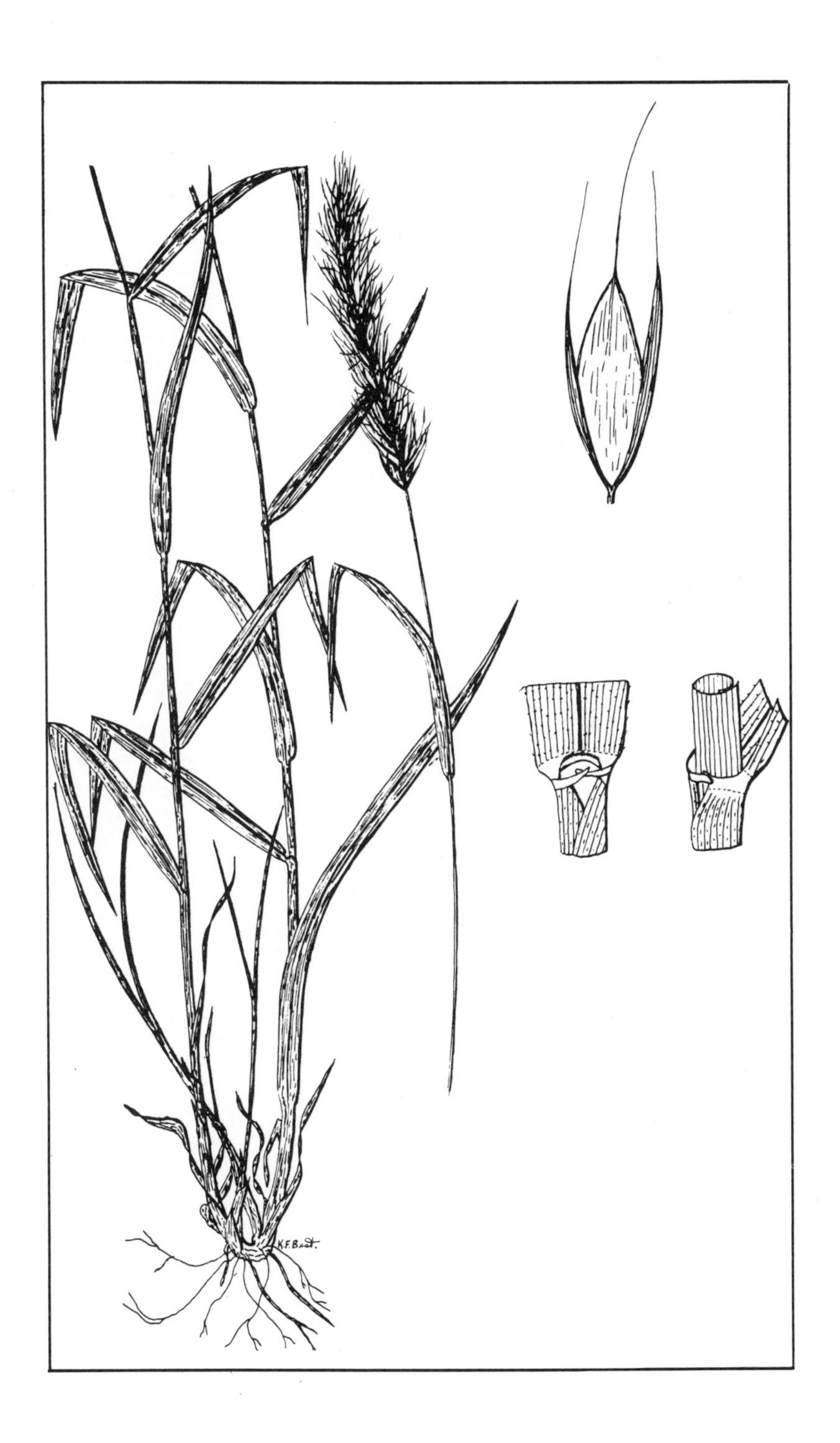

Herbs not otherwise listed

False Solomon's-seal *Smilacina stellata* (L.) Desf.

False, or star-flowered, Solomon's-seal is a member of the lily-of-the-valley family and is the most common of the 15 representatives of that family occurring in the prairie area. It is distributed across Canada and from the boreal forest southward as far as Virginia and Colorado. It seldom occurs in dense stands but rather as spaced plants among the associated grasses. Although it may be found in relatively dry areas, its occurrence is more common in moist meadows and around groves of trees. It is known also as three-flowered Solomon's-seal and wild spikenard.

Smooth, light green stems, 30 cm or more in height, grow from white rootstocks. Six to ten smooth leaves up to 15 cm long grow on opposite sides of the stem, but not in pairs. Small, delicate, white, star-shaped flowers grow at the summit of the stem. Greenish berries with dark stripes in the immature stage turn black at maturity.

Very little is known about the usefulness of this plant except that it is grazed readily by all classes of livestock. Its early spring growth is particularly attractive to sheep and cattle, and even when it is cured sheep have been observed grazing its dry leaves. However, because of its open stand, it does not add much to the forage supply.

Fireweed *Epilobium angustifolium* L.

Fireweed is one of the most common members of the evening primrose family growing in the prairie area. It has a wide distribution throughout Asia and Europe, and it is common in North America from Newfoundland to Alaska and from the northern limits of growth southward through the United States and into Mexico. Its common name, fireweed, is apt because it is one of the first plants to invade fire-blighted forest. It is also called willow herb because its leaves resemble those of willow and its masses of silky seeds are suggestive of willow catkins. In the prairie area it is found around bluffs and sloughs, in open forests, along roadsides, and throughout moist prairie. It is not common in the dry central region.

Fireweed grows from long creeping roots. Its leafy stems may be as much as 1.2 m tall. It has a long flowering season because the lower florets blossom first, followed in succession by those nearer the top of the stem. The narrow leaves become red in the autumn. The long flower stalk bears pink or purple flowers that develop into four-sided pods filled with numerous small seeds, each with a parachute of silky white hairs.

Reports vary considerably regarding the grazing value of fireweed. Sheep make good use of it but apparently require other fodder to balance their rations. Cattle browse it occasionally, whereas horses and game seldom eat it. Fireweed is excellent bee pasture.

Heavy grazing early in the season increases tillering, and subsequently the pasture yield, during the current season. Although this practice eventually reduces the vitality of individual plants, it is one that should be followed in an effort to reduce the fire hazard created by heavy stands.

A.C. Budd

Snakeroot *Sanicula marilandica* L.

Snakeroot is a member of the parsley family and is common in moist woodlands at the edge of the prairie area but is rarely encountered in the drier central portion. It seldom grows in dense stands, but the leafy individual plants add considerable fodder to the associations where it is common.

Snakeroot is a perennial herb. It produces three or four stems that grow from a twisting rootstock. Its leaves are palmately divided, each with from three to five leaflets growing from one point. As with other members of the family, the leaflets have saw-toothed edges. Flower clusters grow in balls about 15 mm in diameter. Seeds are about 6 mm long and are covered with fine, hooked bristles.

This plant is eaten readily by sheep and cattle during its early growth but loses it palatability by the time it flowers. This character is possessed by several members of the parsley family, including musineon, *Musineon* spp.; squawroot, *Perideridia* spp.; cow-parsnip, *Heracleum lanatum* Michx.; water-parsnip, *Sium suave* Walt.; prairie parsley, *Lomatium* spp.; and alexander, *Zizia* spp. Thus the parsleys add considerably to the forage supply at a season before the grasses have completed their growth. In general, all the plants mentioned can be utilized quite heavily, not only because they produce quantities of seed but also because their fleshy roots store reserves of food.

The Indians, also, used the fleshy roots of many native members of the parsley family as sources of food and medicine. The roots were eaten raw, baked, or roasted or were dried and ground into flour for biscuits; the young leaves were collected for greens. The common names of some species and genera, for example biscuitroot and squawroot, suggest the value of these plants for food.

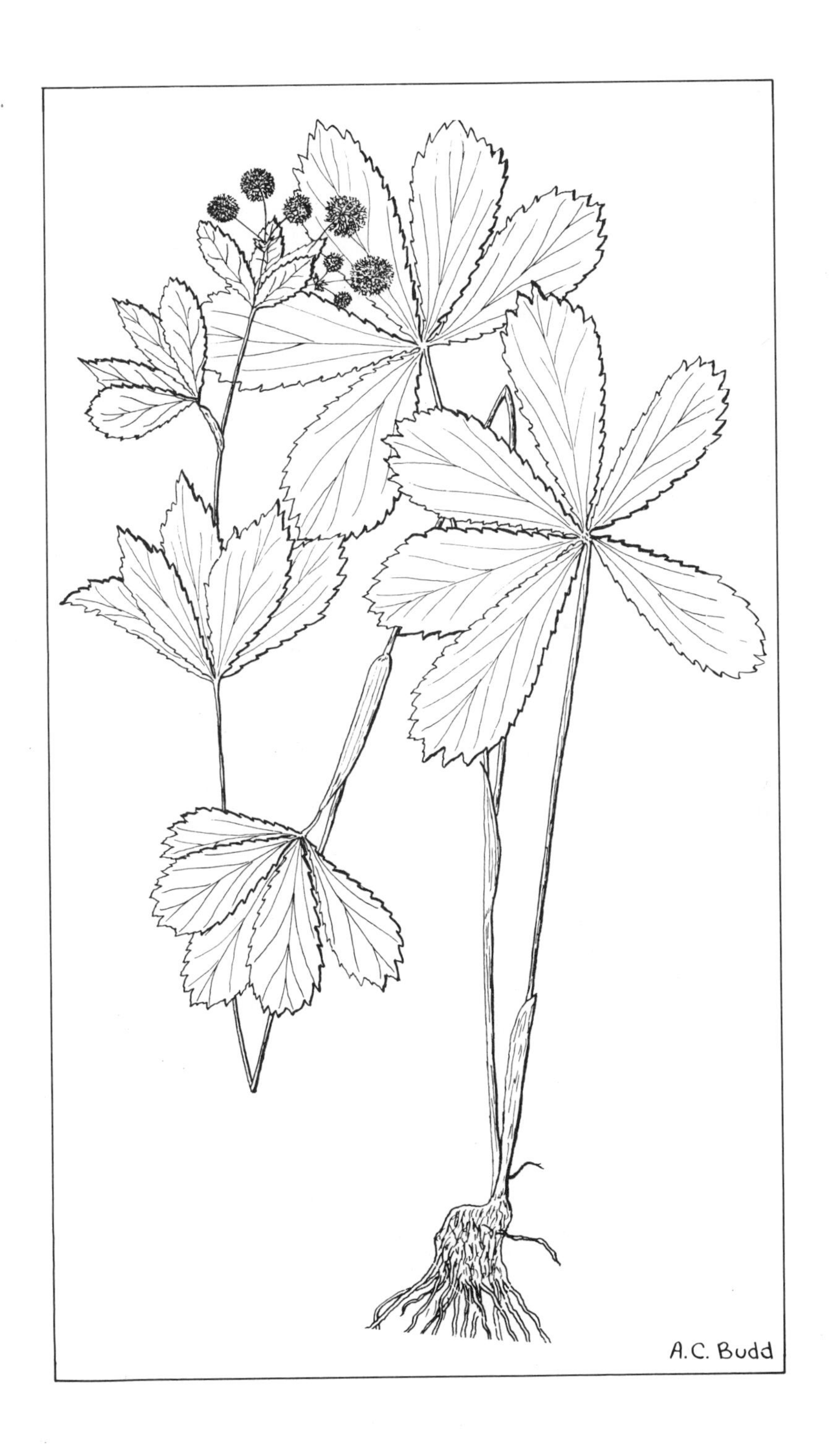
A.C. Budd

Sweet cicelys *Osmorhiza* spp.

The sweet cicely group is another member of the parsley family and is common in the wooded regions at the boundary of the prairie area and throughout the Cypress Hills and Foothills. There are two important species: smooth sweet cicely, *O. aristata* (Thunb.) Mak. & Yabe, which occurs principally in the southern Foothills; and blunt-fruited sweet cicely, *O. chilensis* Hook. & Arn., which may occur anywhere within the region. Sweet cicelys are known also as sweet-anise and sweetroot.

The parsley family is a large one. Besides the sweet cicelys, it includes useful range plants such as snakeroot and cicely, as well as those with poisonous properties such as spotted water-hemlock, *Cicuta maculata* L. var. *angustifolia* Hook. Fortunately, spotted water-hemlock can be identified readily because its several tuber-like roots and its cross-sectioned crown are not possessed by other members of the family. Because the leafage of sweet cicelys can be confused with that of water-parsnip, the root character should be checked to ensure identification.

Sweet cicelys are tall herbs growing from a single, thick, aromatic root. There are three to five leaves on each leaf stem; these leaves are divided into leaflets, each of which is deeply toothed. There are relatively few white or purple flowers, which develop into deeply grooved fruits 1-2 cm long.

All portions of sweet cicely plants are relished by sheep and cattle and are eaten by horses and deer. However, its palatability and nutritive value are destroyed by frost. Because it stores food reserves in its fleshy root, it can be grazed fairly intensively without killing individual plants or overgrazing stands.

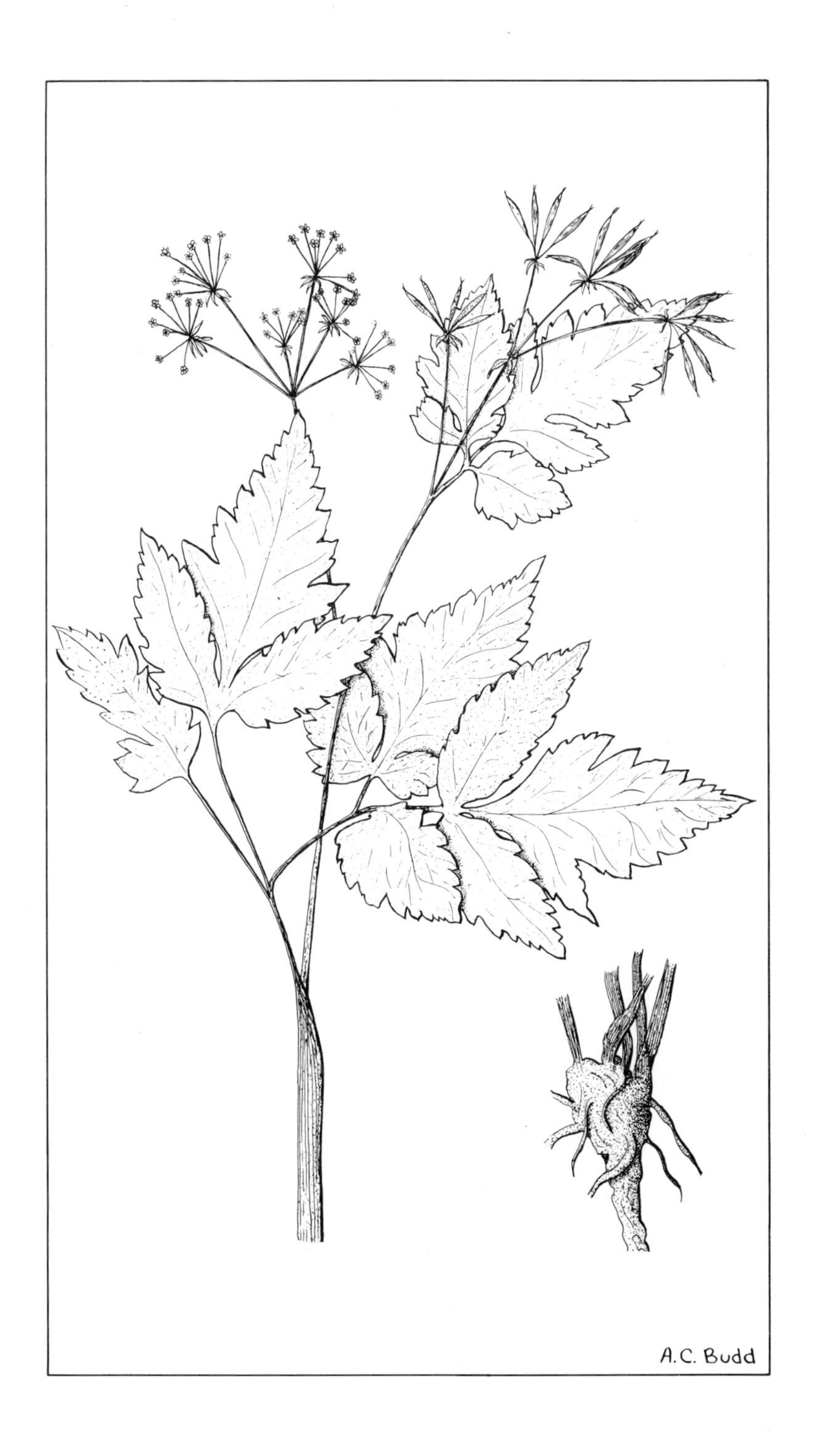
A.C. Budd

Pasture sage *Artemisia frigida* Willd.

The common, low-growing sage of the prairies has been called pasture sage for many years. Probably fringed sage is a more descriptive name because its leaves are small, delicate, and lacelike. It has a wide range, occurring naturally in northern Europe and Asia and throughout western North America. The name *frigida* refers to the frigid regions in Siberia where the original collections were made. It is a member of the wide-ranging composite family.

Pasture sage has a perennial root and crown, but most of its aerial growth is annual. It is a low plant, sometimes extending from a central crown by perennial prostrate stems that root at points of contact with the soil. Its woolly stems seldom exceed 30 cm tall and are crowded with small, silvery leaves. Yellow flowers bloom from late July to early September, and in dense stands they give off clouds of pollen that may cause hay fever.

The forage value of pasture sage as determined by chemical analyses is very high. Protein, phosphorus, and fat contents are well above those for the associated grasses in all stages of growth; in fact, the protein content equals that of good alfalfa. However, the presence of aromatic oils apparently limits its palatability, because only during the autumn and winter do cattle graze the plant to any degree. Sheep eat it more readily and pasture it from early autumn through the winter until late spring.

In many areas the presence of heavy stands indicates an overgrazed condition. Numerous viable seeds, drought tolerance, and low palatability, as well as the ability to root from stems, are all factors that tend to increase stands of pasture sage as the associated grasses are depleted by heavy use. In the mixed-grass prairie area, particularly, increasing stands indicate that overgrazing is being practised.

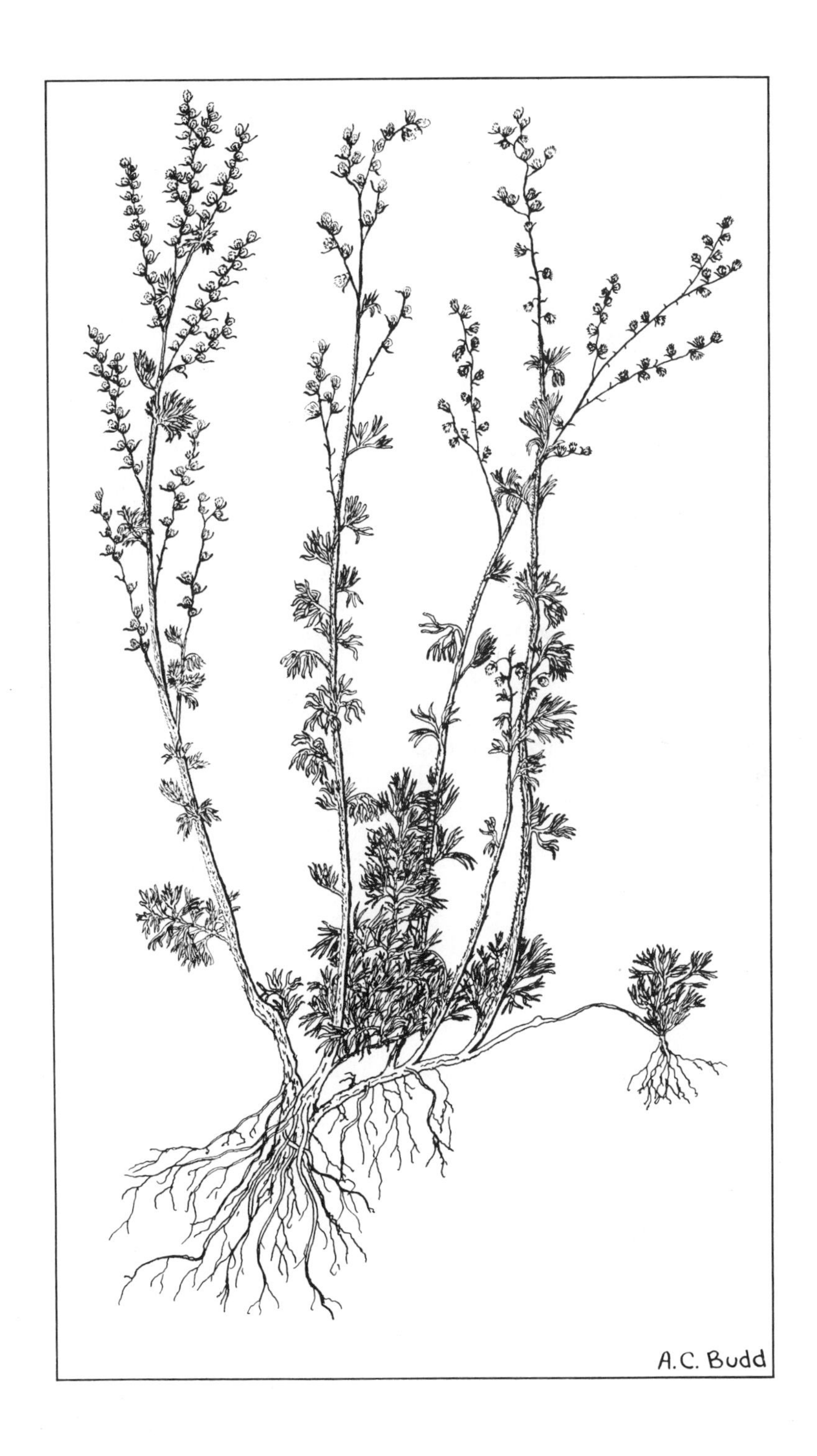
A.C. Budd

Red Indian paintbrush *Castilleja mineata* Dougl.

Red Indian paintbrush, or Indian paintbrush, is the most common of the eight *Castilleja* species growing in the prairie area. It, as well as the other members of the group, occurs as a rather unimportant but colorful member of the open forest communities and grasslands. Paintbrushes seldom grow in a dense stand, but as spaced individuals among grass and shrubs. The paintbrushes are known also as painted cups. They are members of the figwort family.

Semiparasitic in nature and with small root systems, paintbrushes are short, sparsely leaved herbs. The distinctive red, crimson, or yellow portion of the flower cluster is not a part of the flower itself; rather, it comprises small, colored leaves that enclose the true flower parts. All leaves, whether enclosing the stem or the flower, are narrow, without stems, and usually entire.

The figwort family has many members among the range vegetation of the prairie area, several of them quite abundant. However, except for the paintbrushes, none of this numerous family is palatable. Even paintbrushes are not eaten readily by cattle, but sheep graze them throughout the summer.

Paintbrush is a very descriptive name for this group of plants. Their general appearance is that of a brush whose tip has been immersed in red or yellow paint. Because they have a long growing season, they add color to the sward before, during, and after the blooming period of most range flowers.

A.C. Budd

Smartweeds *Polygonum* spp.

Smartweeds are members of the buckwheat family and grow in shallow water or along the margins of nonsaline sloughs throughout the northern hemisphere from the Equator to the Arctic Circle. In the prairie area they dominate the vegetation in sloughs that contain less than 50 cm of water in the spring and that dry up by mid-July. The common codominants are awned and beaked sedges. Smartweeds are known also as lady's-thumb because of the darker leaf center, and as water pepper and persicaria. Both annual and perennial forms are known. The same species may grow on land or in water.

The spikelike cluster of small, white or pink flowers is surrounded by leaves that are long, oval, or lance-shaped and that have darker green centers. Most aquatic forms are hairless, but the leaves and stems of those growing on land are usually very hairy. The stems have joints similar to grasses; the joints are enclosed by small, membranous or leafy sheaths.

The smartweeds that grow in water and that can be harvested as sloughs dry up, such as *P. amphibium* L., water smartweed (illustrated), make excellent aromatic hay; but those that develop entirely on dry land are harsh and unpalatable. Leafy and green smartweed hay contains at least 15% protein with correspondingly low levels of crude fiber. In combination with the sedges or spangletop, smartweeds make palatable and nutritious but light hay. Smartweed is less attractive for pasture and is seldom grazed. In fact, smartweeds that are pastured are credited with causing "yellows" or "big head" in livestock with white skins; the condition is seldom, if ever, produced when cattle are fed well-cured smartweed hay.

K. F. Best

Asters *Aster* spp.

Aster means star and refers to the starlike appearance of the ray florets. Asters probably comprise more species than any other group belonging to the thistle family occurring in the prairie area. More than 25 species have been identified and named. Because of their number, the species vary greatly in type of growth, flower color and size, and palatability.

Asters are usually perennial leafy herbs that grow from thick and often branched rootstocks. Stems may be numerous or solitary, and either herbaceous or woody. Usually there are several flower heads, ranging from less than 5 mm to more than 30 mm in diameter. However, all species have flower heads composed of many florets, with rays that do not number over 50 and at least three rows of bracts that enclose the flower.

The woody asters are not eaten, nor are those herbaceous forms with small white or blue flowers. As they have very definite habitats, usually around sloughs, they do not spread into the adjacent range. Thus they persist but are not useful economically.

Only one native aster can be rated as highly palatable. It is the showy aster, *A. conspicuus* Lindl. (illustrated), named because of its large, violet to blue flower heads often 35 mm in diameter. It has a few large leaves up to 15 cm long and a rough, hairy stem. Not only is it palatable but it has a satisfactory balance, containing as much as 20% crude protein and over 0.50% phosphorus prior to the flowering period. It is quite common throughout the parkland region, Foothills, Cypress Hills, and central British Columbia.

Smooth aster, *A. laevis* L., is palatable also but is not eaten so readily as showy aster. Whereas smooth aster maintains itself and increases slightly with heavy grazing, showy aster is grazed out. Smooth aster has a range similar to that of showy aster.

A.C. Budd

Dandelion *Taraxacum officinale* Web. in Wiggers

Dandelion is likely the best known and the most widespread plant throughout the world. It grows on a variety of soils but prefers moist meadows; it grows under a wide range of climate, although it prefers moderate conditions. It has spread from Europe throughout the settled portions of both the northern and southern hemispheres. The name dandelion means lion's tooth and refers to the deeply cut and backward-pointing segments of the leaves.

Dandelion grows from a deep, fleshy taproot. There is no stem and both the numerous leaves and the flower stalks grow directly from the crown. The leaves are deeply cut; they may stand erect when growing in shady places or cling to the ground when growing among short grass. A single yellow flower head grows at the tip of the hollow flower stalk. Both the leaves and the flower stalk contain an acrid, milky sap.

Dandelion is palatable to all classes of livestock and is particularly so to sheep. Because dandelion makes an early growth and continues to produce leaves throughout the summer until late fall, this plant is one of the few herbs that provide forage throughout the growing season. Its abundance may be an indicator of overgrazing in certain areas, but it might also indicate a lack of mineral balance in the soil. Dandelion is considered to be one of the best bee pastures during the early summer, one of the good characters that help to offset its reputation as a weed of lawns, pastures, and irrigated hayfields.

A close relative to the common dandelion is the red-seeded dandelion, *T. erythrospermum* Andrz. It differs principally in leaf shape and color of seed. Its leaf has a small end segment and its seeds are red, whereas common dandelion leaves have large end segments and its seeds are brownish. Both occur everywhere throughout the prairie area and both are quite resistant to grazing.

A.C.Budd

Groundsels *Senecio* spp.

The genus *Senecio* is one of the largest groups of plants known, with more than 2600 species being reported from various places throughout the world. These species show a great variety of growth forms, from giant herbs and shrubs and moderate-sized trees in the equatorial zone of Africa to small herbs in the mountain area of Western Canada. Some 15 species occur in the prairie area. Of these, three or four are palatable and one, an invader from Europe, is reported to be poisonous to cattle and horses.

Although the groundsels, also named ragworts or butterweeds, superficially resemble many other members of the thistle family, the single row of bracts that enclose the flower head is a character that distinguishes this group from others. These bracts are strap shaped, with a lance-shaped tip. The leaves are alternate, usually entire or sparsely saw toothed, although those of the reputedly poisonous species, tansy ragwort, *S. jacobaea* L., are deeply dissected into three or four segments.

One or more of the groundsels may be found anywhere in the prairie area. One of the most common is marsh ragwort, *S. congestus* (R. Br.) DC., which often forms dense communities around sloughs and lakes and along streams; it is fairly palatable to sheep but is rated as poor fodder for cattle. Silvery groundsel, *S. canus* Hook., is the most common species in the dry central portion. It is seldom grazed by cattle, but sheep eat it readily. The poisonous species, tansy ragwort, has been reported at the eastern edge of the prairie area in Swan River, Erickson, and Winnipeg. It is palatable, but cattle feeding on it may develop a condition known as Pictou disease. It has not become very common, except locally in moist areas.

a.C.Budd

Hawk's-beards *Crepis* spp.

There are six members of the hawk's-beard group found in the prairie area. They have the typical flower head composed of numerous florets. The milky sap is a character it has in common with dandelion. None of the species occurs throughout the region, each having a fairly local distribution or requiring specific growth conditions for its expression.

The native hawk's-beards are perennial herbs with deeply penetrating, woody taproots. The toothed leaves are mostly basal, although a few much smaller leaves may occur on stems 25–50 cm tall. Both leaves and stems are hairy but not woolly. The flowers are yellow in all species, usually occur in groups of three to five, and are about 25 mm across. A European annual has become widespread in the parkland area; it differs from the native species in that its flower heads are smaller and more numerous.

Hawk's-beards are palatable to all classes of livestock. Sheep tend to overgraze stands to the point where all the plants are killed out. Cattle make fair to good use of them and horses eat them readily. Deer and elk are reported to eat them almost to the same degree as do sheep.

Unfortunately, hawk's-beards do not grow in dense stands. They appear instead as scattered plants among the associated vegetation. Although they do not produce an abundant supply of forage, their fairly dense basal leafage assures a good bite for the grazing animal. Their principal habitat is open woods and meadows at the northern boundary of the region, but one species, which grows in moist, saline flats, may be found anywhere within the prairie area.

A.C. Budd

Perennial sow-thistle *Sonchus arvensis* L.

Perennial sow-thistle, or sow-thistle, is an introduced weed coming originally from Europe. It found many sites where growth conditions favored its establishment, and its aggressive and rapidly spreading characters soon caused it to be recognized as an important noxious weed of cropland. In the prairie area it spread rapidly through the Portage Plains and Red River Valley before 1920 and more slowly westward throughout the parkbelt. It moved into the drier portion of the region during the late 1930s and established itself in small patches in favorable locations.

Perennial sow-thistle grows from vigorous, creeping, white rootstocks that extend 10–50 cm below the surface of the ground. Hollow stems up to 1.5 m tall have large and numerous leaves, which have lobes that point backward. The showy, yellow flowers may reach as much as 5 cm in diameter and grow in clusters at the top of the stem. An acrid, milky juice can be squeezed from all parts of the plant.

As with many other members of this family, perennial sow-thistle is grazed readily, provided the stands are not too large. Individual plants or even small patches may be eaten but large stands are not touched to any degree. However, it is an early-growing plant and thus often adds to forage supplies before associated grasses come into strong production. Perennial sow-thistle has a marked laxative effect on cattle.

A.C. Budd

Umbellate hawkweed *Hieracium umbellatum* L.

Umbellate hawkweed, or common hawkweed, is the most abundant species of the five hawkweeds that occur within the prairie area. Not only is it the most abundant but it is also the most widespread, as all other species have local distributions. The generic name *Hieracium* is derived from the Greek word meaning hawk.

Umbellate hawkweed grows from a stout rootstock. Its stem is unbranched except at the top, where short branches end in flower clusters. The leaves are sparsely toothed, with their bases clinging to the stem. Both stem and leaves contain a milky sap and both are covered with dense, short, white hairs, giving the plant a woolly appearance. Flower heads are yellow, about 25 mm across, and contain ray florets with a square-toothed tip.

All livestock graze umbellate hawkweed, although sheep utilize it the most and horses the least. All parts of the plant are eaten. Although the leaves and flowers are the most palatable to domestic livestock, deer and elk also graze all parts readily.

Other native hawkweeds resemble umbellate hawkweed in most characters. All species have a woolly appearance, all have cream-colored or yellow flowers, and all have entire or very sparingly toothed leaves. None of these species are abundant and none have widespread distributions.

G.C.Budd

Legumes

Alfalfa *Medicago sativa* L.

Alfalfa, or lucerne, spread throughout Europe and Asia from its home in Asia Minor prior to and following the beginning of the Christian era. It was introduced into North America early in the 18th century but its value was not recognized until after 1850. The variety Grimm was selected from seed imported from Germany, whereas the Ladak variety was developed from seed obtained in northern India. Siberian alfalfa, *M. sativa* ssp. *falcata* (L.) Arcang., was imported from central Siberia. It is the most tolerant of cold and drought of the several species comprising the group.

Alfalfa is a deep-rooted herb with a woody crown that grows immediately above or below the surface of the ground. It has numerous three-leaflet leaves crowding stems 30–80 cm tall. The purple flowers of *M. sativa* develop into coiled pods containing up to 10 seeds, whereas the yellow flowers of *M. sativa* ssp. *falcata* grow into easily split, sickle-shaped pods with only three or four seeds.

Alfalfa is palatable to all livestock and is considered one of the best pasture and hay plants throughout the temperate zones of the world. Its digestibility when fed alone is not high, seldom over 55% on a dry-matter basis, but its high protein and mineral contents make it an excellent supplementary feed to balance rations. When mixed with native hays, it increases their digestibility and apparent palatability. When grown in mixtures with grasses, it lengthens the life of the stand; the mixture out-yields either grass or alfalfa grown alone. It is the best legume crop for irrigated land and has produced excellently on dry land.

Several improved varieties and strains have been developed. Disease resistance, drought tolerance, winterhardiness, and creeping roots that provide better grazing tolerance have all been bred into high-producing varieties, and selection can be made from this material for pasture and hay.

Alfalfa may cause bloat when pastured. However, grass–alfalfa mixtures are less dangerous than is alfalfa alone, dryland pastures are less hazardous than irrigated fields, and continuous grazing is less risky than rotation.

A.C. Budd

American hedysarum

Hedysarum alpinum L.
var. *americanum* Michx.

Hedysarum species occur throughout the north temperate zone. American hedysarum, also known as sweet brome or sweet vetch, is abundant in the Foothills, the Cypress Hills, Wood Mountain, and the parklands and is the only species common thoughout the Prairie Provinces. It occurs in fescue prairie and at margins of woods.

American hedysarum is a fairly erect plant. Its compound leaves may bear up to 10 pairs of leaflets. The flowers range from white to purple and grow in racemes at the ends of the erect stems. The flat seed pods are unique; there is a constriction between adjacent seeds and the pods break between the seeds rather than splitting along the back. Conspicuous veins run across the seed pod.

Reports on its nutritive value vary considerably, but mistaken identification is one cause for the variation. In general *Hedysarum* species are palatable to sheep and fairly so for cattle during midsummer and early autumn. Overuse of these species usually indicates an overgrazed range.

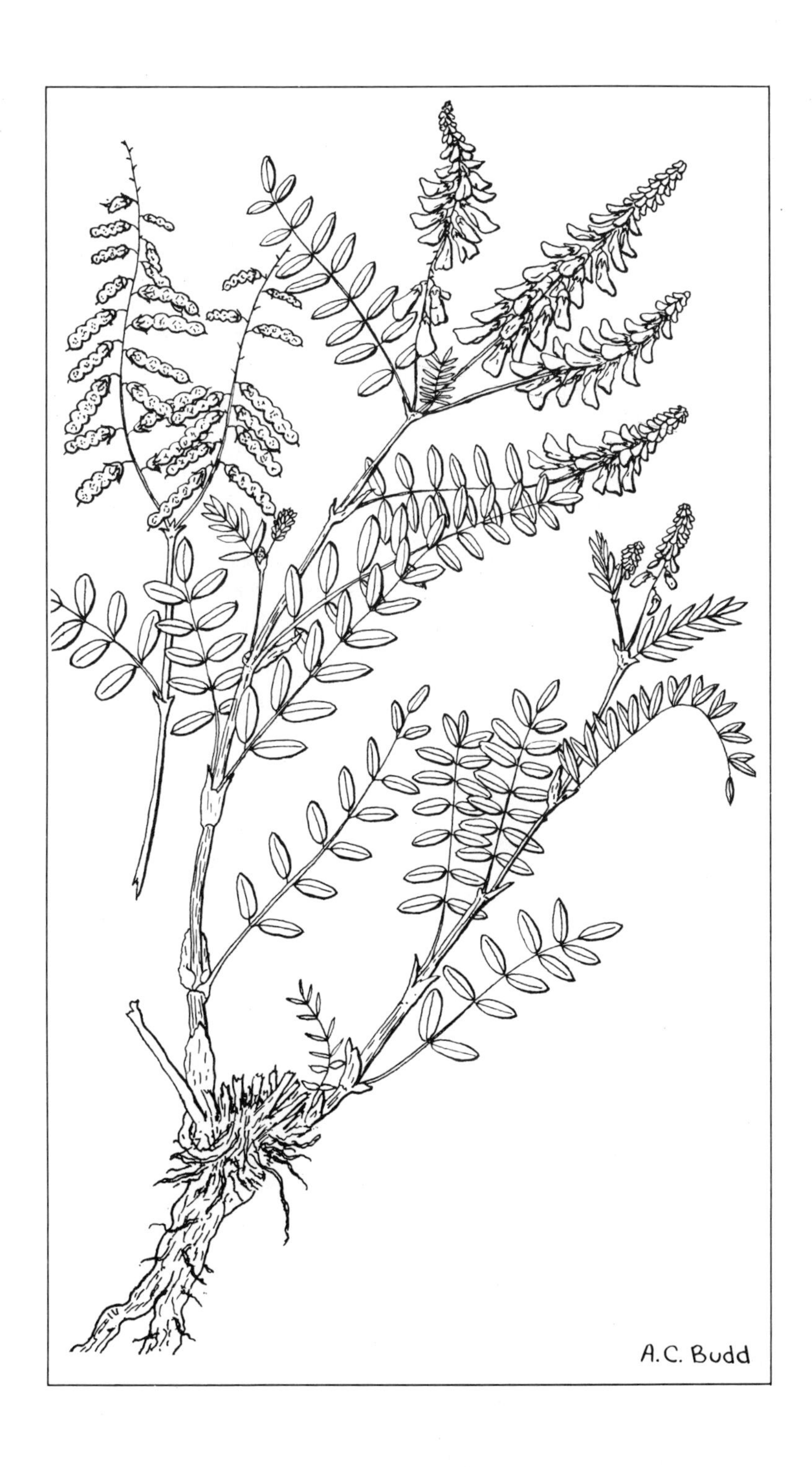
A.C. Budd

Bird's-foot trefoil *Lotus corniculatus* L.

Bird's-foot trefoil was introduced into North America from Europe, probably in the late 19th century. In Europe, the species is widespread and used as forage in many countries. As a native plant it occurs in moist steppes and meadows.

Bird's-foot trefoil is a perennial legume with a well-developed taproot and several stems up to 80 cm tall, arising from the rootcrown. Stems are slender and moderately leafy. The flower heads have one to five yellow flowers in an umbel at the tip of a long, slender peduncle. The pods are set at a right angle, and several together resemble the toes of a bird's foot, hence the common name.

In the prairie area bird's-foot trefoil is adapted to conditions where precipitation exceeds 400 mm. A very similar species, *L. pedunculatus* Cav., has been introduced with seed of bird's-foot trefoil in some areas. It has hollow stems and has up to 10 flowers on each umbel. Trefoils do not cause bloat. They are highly palatable and nutritious, with a protein content that may reach 28% in May and is still around 19% in July.

R.M.B.

Clovers *Trifolium* spp.

The true clovers form a large group of the legume family. A few are native to Canada, but those of economic importance have been introduced from Europe where they have been cultivated for several hundred years. The species of agricultural importance in the prairie area are white clover, *Trifolium repens* L. (illustrated, *bottom*); alsike clover, *T. hybridum* L. (illustrated, *top*); and red clover, *T. pratense* L. Each of these species has named varieties that are adapted to certain districts or soils.

White clover has creeping stems that root and produce new plants, but neither the red nor alsike clover has this character; both grow from crowns. All species produce three-leaflet leaves that have finely toothed margins. Flowers are borne in crowded heads and develop small pods hidden in the dry flower parts. Most true clovers are perennials but a few species of little economic importance are annuals or winter annuals.

Clovers generally are palatable to all classes of livestock. Their high protein content make them valuable supplements to poorer-quality feed. Bloat may occur in animals grazing on particularly lush pastures; white clover is the most dangerous. Clover leaves have reportedly been used as a human food, particularly in salads, and dried flowers and seeds have been made into flour for pastry and bread. Most of the true clovers produce good bee pasture.

Clovers have considerable economic importance for pasture purposes at the margin of the prairie area and for irrigated pasture within the drier sections of the region. They respond well in districts where the rainfall is greater than 400 mm a year and where the soil contains satisfactory levels of phosphorus and calcium. Red and alsike clovers grow particularly well in the boreal forest zone.

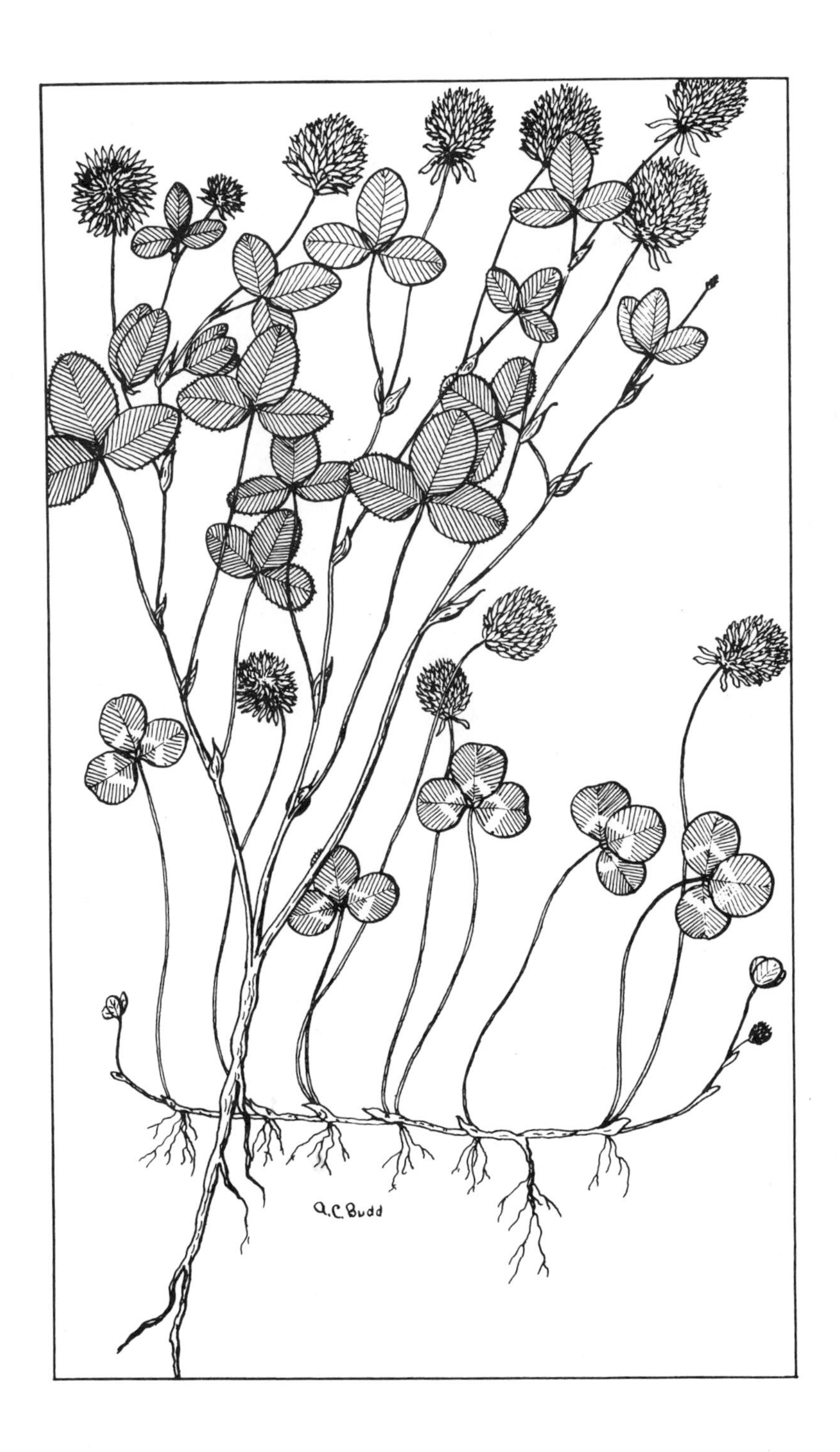
a.C.Budd

Cicer milk-vetch *Astragalus cicer* L.

Over 20 species of the *Astragalus* or milk-vetch group of the legume family occur in the prairie area. Among these are species poisonous to livestock, as well as a few that are apparently palatable and nutritious.

Cicer milk-vetch has been introduced from Europe, where it is widespread in grassland and open woods in the southern half of the continent. Several leafy stems, growing to 1 m high, arise from stout, creeping roots. The leaves have 10–15 pairs of sparsely hairy leaflets. Flower heads are borne at the tips of the stems, as well as on peduncles arising from the leaf axils, and have as many as 20 yellow flowers. The pods are ovoid, inflated, and covered with white and black hairs.

Cicer milk-vetch is palatable when young and is used for hay in some districts under irrigation.

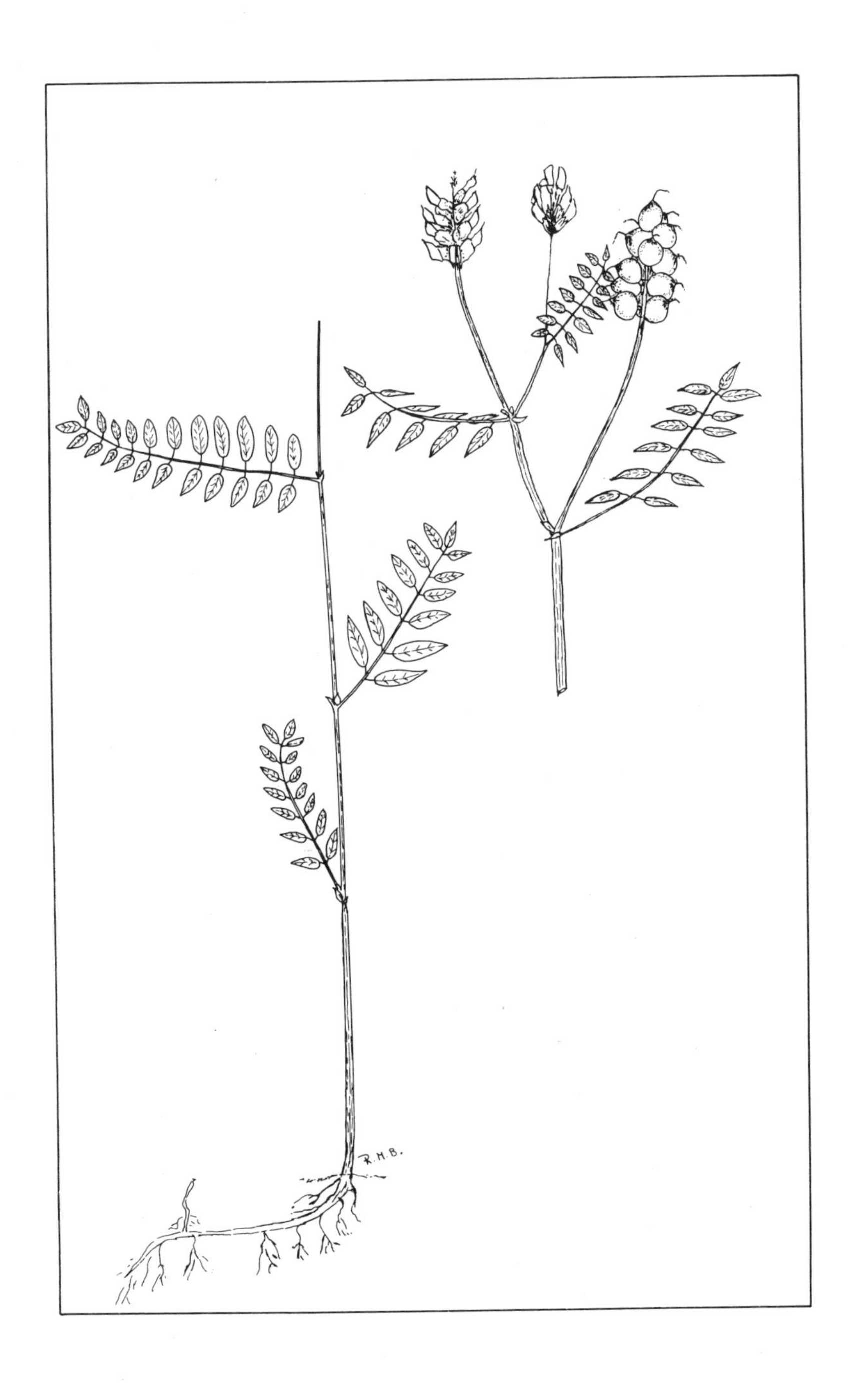
R.H.B.

Ground-plum *Astragalus crassicarpus* Nutt.

Among the most interesting of the *Astragalus* spp. is the ground-plum. This species clings to the ground, the stems creep between grass plants, and the leaves intermingle with the grass leaves. Single plants may be as much as 1 m in diameter. Small, creamy, purple-tipped or bluish purple flowers develop into plum-shaped fruits up to 25 mm long. The fruits are green before maturity but become pale brown or even reddish when ripe. The Indians of the prairie area reportedly ate the green fruit either raw or boiled, and cattle have been observed to search pastures for both fruit and leafage.

Ground-plum is becoming rare. It prefers the deep, fertile soils selected by homesteaders as the best farm land and has been heavily grazed. Although there are few records indicating its nutritive qualities, those available suggest relatively high protein and low fiber contents.

In the native prairie other *Astragalus* spp., namely *A. danicus* Retz., purple milk-vetch, and *A. striatus* Nutt., ascending purple milk-vetch, are grazed to some extent. In the parklands *A. frigidus* (L.) Gray, American milk-vetch, is grazed after purple vetchling and palatable grasses have been grazed out.

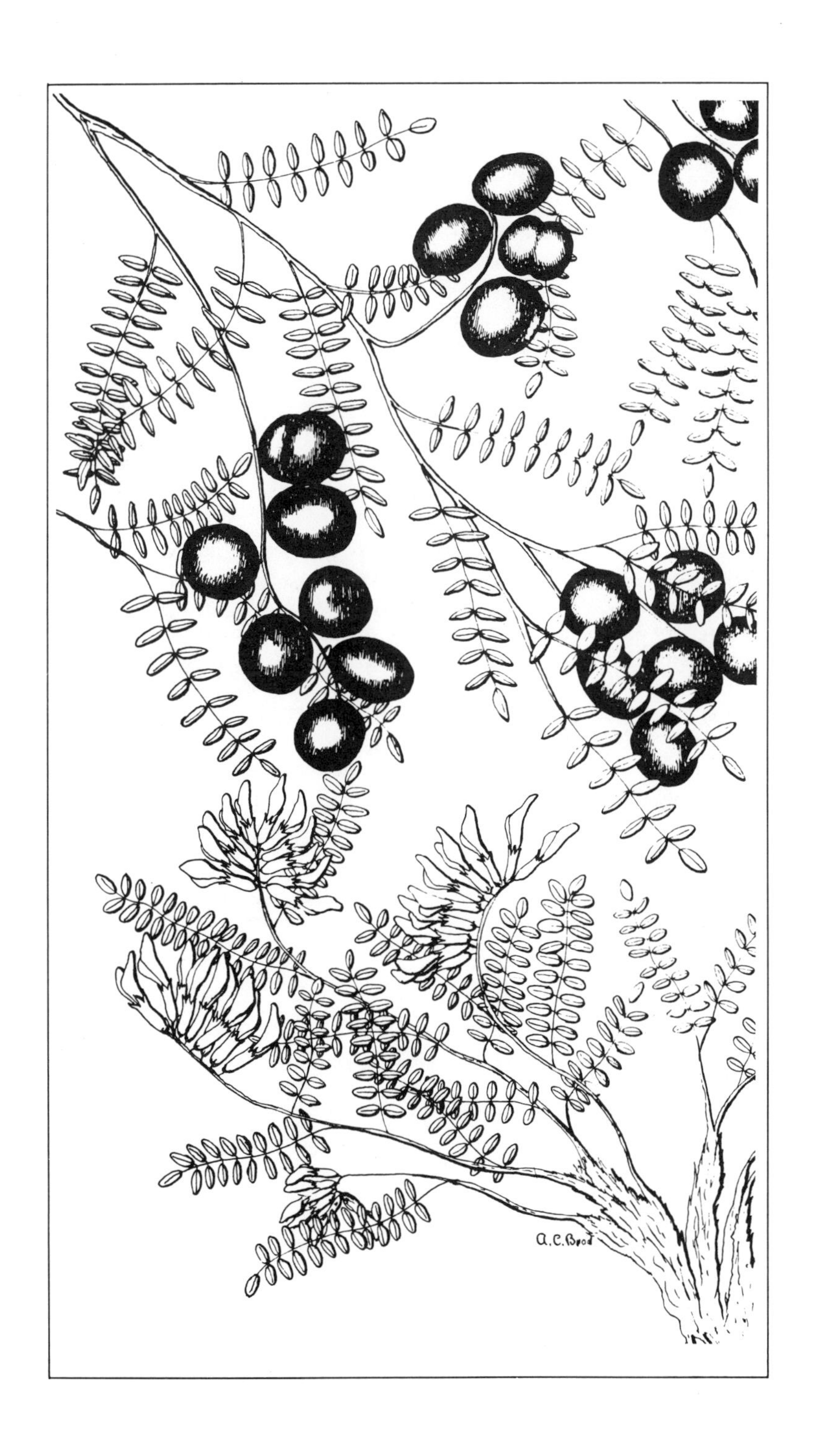
A.C.Brod

Purple vetchling *Lathyrus venosus* Muhlenb.

Purple vetchling, or peavine, is the most common of the four *Lathyrus* species that grow in the prairie area. It has a wide range, growing from Quebec to British Columbia and southward as far as Georgia, Louisiana, and Texas. Purple vetchling is a legume of bushland and forest, where it often dominates the understory. The other three native species likewise occur in woodlands but may also be found on the shores of lakes and marshes throughout the open prairie.

Purple vetchling is a climbing plant with four-angled stems growing from a woody root. Its leaves have four to six pairs of smooth, oval leaflets; a tendril replaces the usual leaflet at the tip. The 12–16 purple flowers grow in a fairly dense raceme and develop into flat, veiny pods up to 5 cm long. One native species, *L. ochroleucus* Hook., has cream-colored flowers; it is commonly found in bushland but may occur in the southern part of the prairie area.

Purple vetchling is palatable. Cattle, sheep, and horses seek it out and fatten on its lush foliage and abundant seed. Unfortunately, the plant does not cure and, as frost destroys its nutritive qualities, it should be grazed before the late summer. Because of its palatability it is easily over-grazed, and as its stands disappear the value of bush pasture is reduced accordingly. Fortunately, the woody root persists for many years, and when bush pastures are rested, purple vetchling often makes a marvellous recovery. No management plan has been worked out to maintain this valuable native forage, but grazing should never be heavy enough to graze off all the seed pods. Nutritive qualities as determined by chemical analyses are excellent, with as much as 30% crude protein in the young leafage.

A.C. Budd

Sainfoin *Onobrychis viciifolia* Scop.

Sainfoin is a legume introduced from Europe, where it is possibly native in central eastern countries. It has been cultivated for several centuries, and its origin is uncertain. Sainfoin usually has several leafy stems arising from a taproot and may reach a height of 80 cm. The leaves have 6–14 pairs of leaflets, which are more or less hairy. Flower heads grow up to 15 cm long and have up to 40 purple flowers. The pod is less than 1 cm long, hairy, and toothed on the margins.

Sainfoin is palatable, and selections adapted to the moister districts in the prairie area are available. It does not cause bloat and makes excellent bee pasture. Protein content in May and July can be as high as 26%.

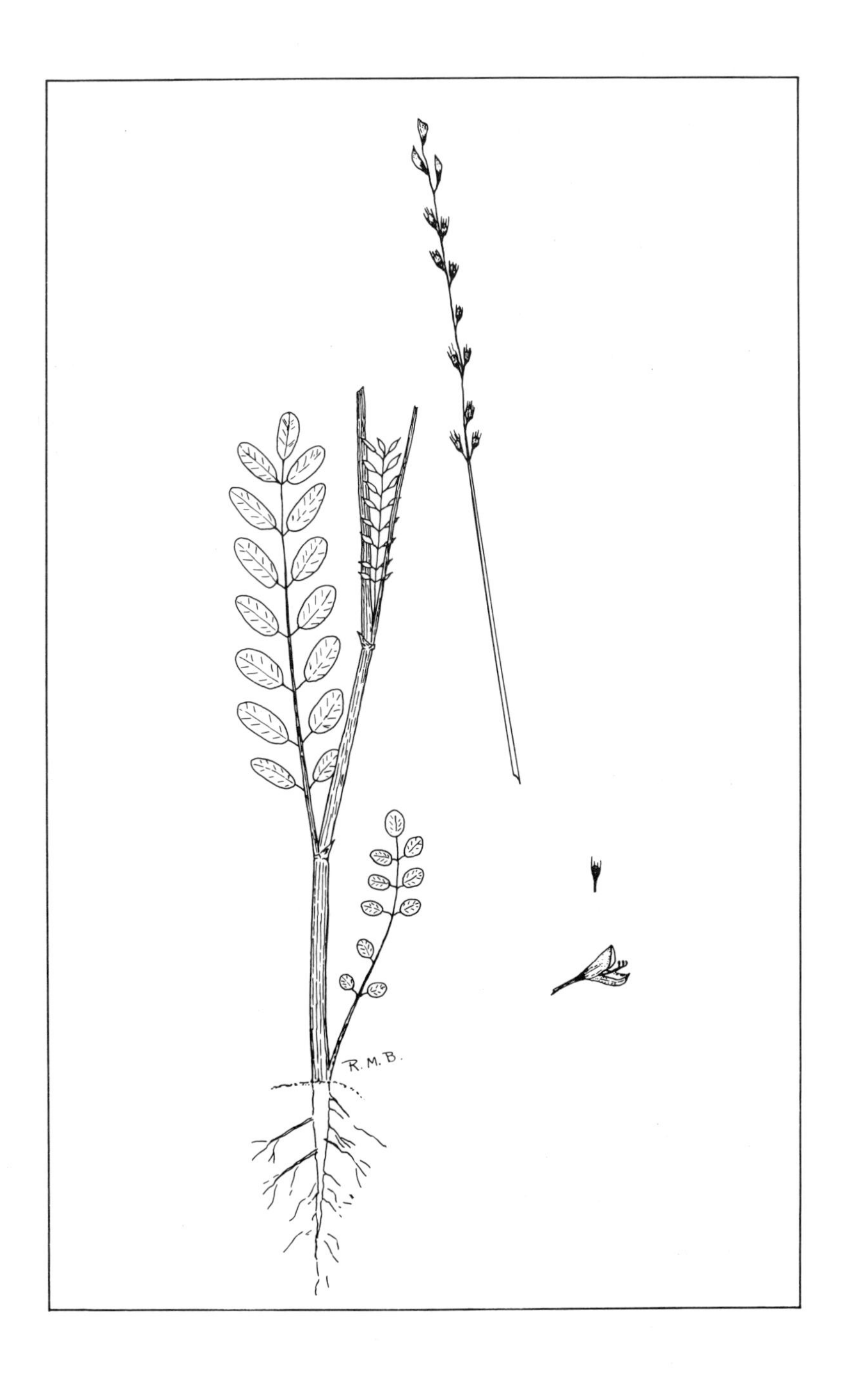
R.M.B.

Silvery lupine *Lupinus argenteus* Pursh

Lupines, which are showy members of the legume family, have palmately divided leaves and long, brightly colored flower spikes. Five species are native to the prairie area, but only the silvery lupine is widespread and of importance as a forage plant. The scientific name *Lupinus* is derived from the Latin word meaning wolf, and one of its many common names is wolfbean. In the prairie area, silvery lupine occurs only in the Cypress Hills, the Milk River bench, and the Foothills about as far north as Rocky Mountain House.

Silvery lupine grows from a persistent woody root. The several stems reach a height of 80 cm and are covered with short hairs. Leaves have six to ten leaflets, all growing from a single point; each of these may be over 5 cm long and are usually hairy. The long racemes bear many white to violet or rose flowers, from which hairy pods containing five or six seeds develop. The woody root persists in cultivated fields; even clean summerfallow sometimes still has several lupines per hectare.

Opinion is divided as to the forage value of silvery lupine. In the intermountain states it is classed as poisonous to sheep and toxic to cattle and horses when grazed; because it is quite palatable it can with reason be classed as a menace in that area. However, no reports of livestock poisoning are available to indicate that it has poisonous properties on prairie pastures; in fact it is considered to be an excellent constituent of upland hay in the Foothills.

A.C. Budd

Sweet-clovers *Melilotus* spp.

Sweet-clover, a native of Asia Minor, was introduced into North America in the early 1700s. It was considered to be a weed for nearly 200 years and only since 1875 has it become recognized as a valuable forage plant. As with most members of the legume family, the roots of sweet-clover produce nodules that extract nitrogen from the air for the use of the plant.

Sweet-clover is a sweet-scented, upright annual or biennial legume. It produces three-leaflet leaves that are saw toothed around most of their edges. Flower racemes grow from the axils of the leaves and produce many white or yellow florets, each of which may develop a seed pod usually containing one seed.

There are two common sweet-clover varieties: white, *M. albus* Medic., and yellow, *M. officinalis* (L.) Lam. They can be distinguished by the yellow flowers, the shorter and finer stems, more spreading growth, finer leaves, and the purple-flecked seed of the yellow-flowered species. Both of these species are biennials, producing seed during their 2nd year of growth.

Sweet-clover has a well-balanced nutrient content during its early growth, but loses much of its protein as the plants mature. In its young stages of growth it provides palatable pasture for all livestock and when cut early and cured it makes a desirable hay. Its high coumarin content quite often prevents stock from eating it until the animals become accustomed to the somewhat bitter flavor. Spoiled sweet-clover hay may be harmful to livestock because decomposition of coumarin results in a toxic compound that causes bleeding. In many districts sweet-clover has been used to control the growth of noxious weeds and to build up organic matter content of soils. Sweet-clover is recognized as one of the best bee pastures available.

A.C. Budd

Vetches *Vicia* spp.

American vetch, *Vicia americana* Muhlenb. (illustrated), is the only native vetch in the prairie area. One variety occurs in moist grassland, and it is low growing; another variety is common in shrubbery and tree groves and can reach 80 cm tall. A few species have been introduced and have locally escaped from cultivation. Of these, *V. cracca* L., tufted vetch, is the most common and can be found throughout the parklands and boreal forest areas.

Vetches are weak-stemmed legumes that often appear to grow as vines. Their four-angled stems are easily broken. Leaves consist of six to twelve or more pairs of leaflets with a tendril or tendrils growing at the ends of the leaves. The flower stalks grow from the axil of the leaves and produce a few flowers at their tips. Seed pods are flat and break open easily. The leaflets of the narrow-leaved American vetch, *V. americana* var. *minor* Hook., are narrow and spear-shaped at each end; those of the American vetch are broadly oval and have a tip at the end.

Vetches have characters similar to many other legumes. On their roots they produce nodules that contain bacteria that fix atmospheric nitrogen, they have the characteristic flower and pod, they are useful for green-manuring, and they provide fair to good bee pasture.

Vetches are palatable and are grazed well by cattle, although not as well as other legumes are. They contain about 20% protein in summer and are low in crude fiber. Like purple vetchling, vetches have a low grazing tolerance and are easily grazed out.

A.C. Budd

Poisonous plants

Arrow-grasses *Triglochin* spp.

Two species of arrow-grass occur in the prairie area. The more common of the two is seaside arrow-grass, *T. maritima* L., which occurs in brackish to highly saline areas, often covering as much as a hectare or even more. Individual plants may be 90–100 cm tall, with leaves 30–40 cm long. Marsh arrow-grass, *T. palustris* L., is much smaller; its stems are seldom more than 50 cm tall, with leaves reaching only up to 20 cm long. It is found along creeks and streams and in marshy areas where the soil is high in calcium and magnesium.

Both species are grasslike, but the leaves are spongy and crescent shaped in cross section; the stems are solid.

Arrow-grasses are very poisonous to all livestock and are the main cause of cattle losses in the prairie area. The foliage is readily eaten, especially when the animals are hungry, thirsty, or in need of salt. Poisoning is most common after cattle have been herded into fresh pasture where arrow-grasses occur.

The toxic principle is prussic acid, which develops in the plant after injury. Grazing and trampling can cause development of the poison, but prussic acid can also form as a result of drought or frost. The first killing frost in the fall can result in poisonings in pastures where cattle had grazed earlier without any harm.

Death camas *Zygadenus gramineus* Rydb.

Death camas is a member of the lily family. Its foliage and flowering stem arise from a bulb about 10 cm deep in the ground. The leaves are grasslike but are thicker and commonly V-shaped, growing up to 15 cm long; the single stem grows up to 50 cm tall and bears an elongated raceme with many small, yellowish flowers.

The foliage of death camas appears early in spring, usually before the grasses commence growing, and thus are often the first grazeable greens available. Death camas occurs in draws, low-lying areas, and lower coulee slopes from south central Saskatchewan west throughout southern Alberta.

All parts of death camas are poisonous, especially the bulb. Thus the plant is very dangerous in spring, when the soil is wet and the bulb can easily be pulled up with the foliage. Although death camas is most dangerous for sheep, cattle can be affected also, particularly when the plant occurs in almost solid stands.

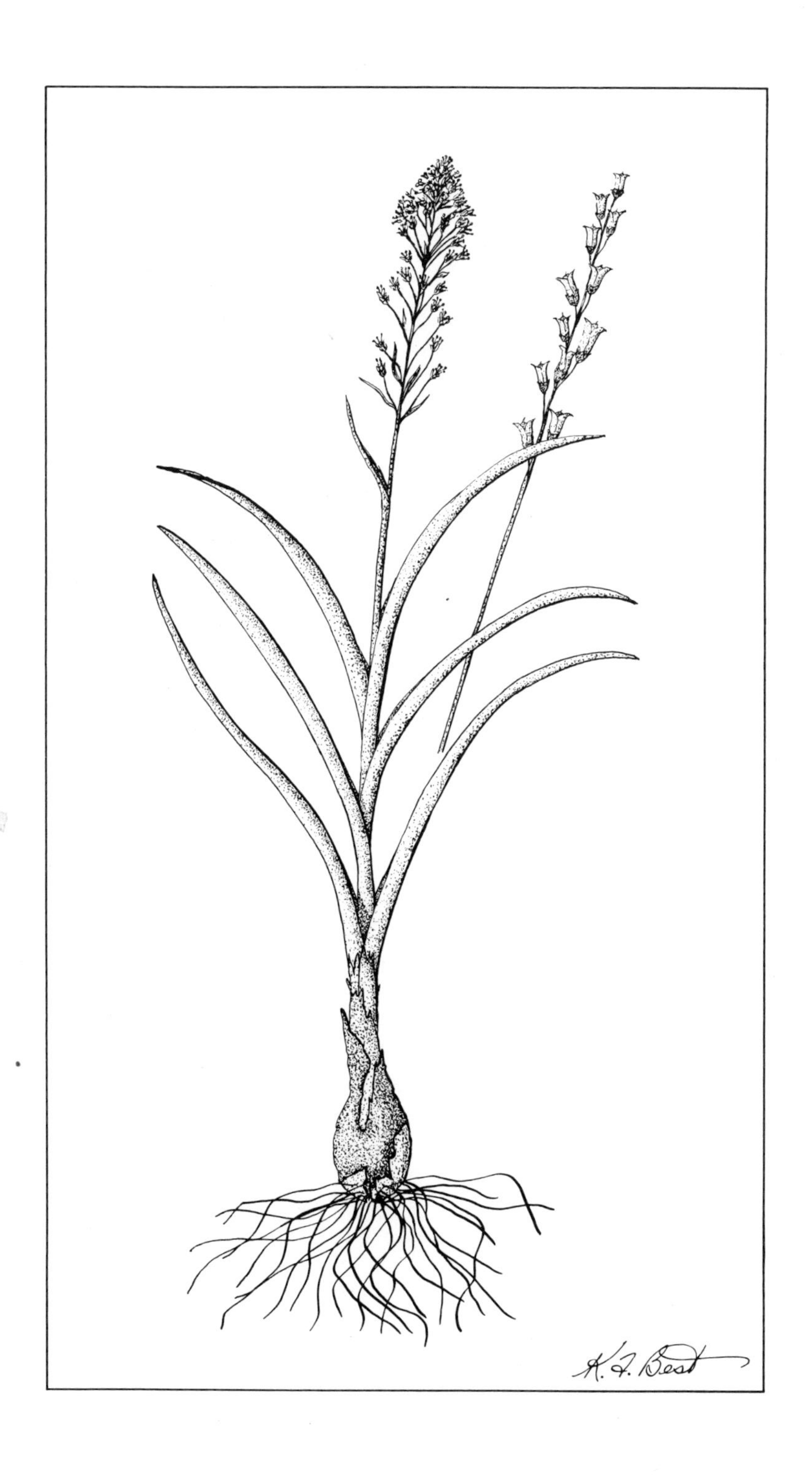

Larkspurs *Delphinium* spp.

Two larkspurs occur in the Prairie Provinces. Low larkspur, *D. bicolor* Nutt. (illustrated, *top*), occurs in the Wood Mountain and Cypress Hills regions of southern Saskatchewan and Alberta and in the southern Rocky Mountains. It grows up to 50 cm tall and has much dissected leaves that are round in outline and extend up to 6 cm across. The few large, dark blue flowers are in a loose, terminal raceme. Tall larkspur, *D. glaucum* S. Wats. (illustrated, *bottom*), occurs in the Rocky Mountains, the northern parklands, and the boreal forest of Alberta and western Saskatchewan. It grows up to 1.5–2 m tall; the leaves can reach 20 cm across and are much dissected, and the blue flowers are borne in dense, elongated, spike-like racemes.

Low larkspur occurs in draws and coulees in moist grassland and shrubbery, as well as at the margins of aspen groves; tall larkspur occurs in open forest, shrubbery, and meadows. Both species are very poisonous to cattle and have caused considerable losses. Low larkspur is especially toxic in spring, when its emerging leaves are often the only green foliage available for grazing. The plant dies off at maturity in early July. Tall larkspur forms considerable lush foliage until late June or early July, when the flowering stems are developed. It is poisonous until fall but is less palatable after the flowering stems have formed.

A.C. Budd

Locoweeds *Oxytropis* spp.

Loco disease can be caused not only by species of *Oxytropis*, the locoweeds, but also by several species of *Astragalus*, the milk-vetches. None of the loco-causing milk-vetches occurs in Canada, but four species of locoweed causing the disease can be found in the prairie area. Of these, early yellow locoweed, *O. sericea* Nutt. var. *spicata* (Hook.) Barneby, and late yellow locoweed, *O. campestris* (L.) DC. ssp. *gracilis* (Nels.) Boivin, are common in dry and moist prairie, respectively. Two other species, Bessey's locoweed, *O. besseyi* (Rydb.) Blankinsh., and purple locoweed, *O. lambertii* Pursh, are less common but can be locally abundant. Both are purple flowered.

The toxic principle, locoine, has not yet been identified as to its exact nature. It is a slow-working poison, and locoweeds must be eaten over periods of two or more months before symptoms appear. Under normal conditions, when ample forage is available, poisoning is unlikely to occur because locoweeds are not palatable. On overgrazed range, however, animals may eat the locoweeds and develop a craving for the plant, actively seeking it out even when more palatable forage is available.

Milk-vetches *Astragalus* spp.

Most of the milk-vetches occurring in the prairie area are not palatable, but none is actually poisonous. At least two species are potentially poisonous, however, through accumulation of selenium. Both these species are common in the prairie area.

Narrow-leaved milk-vetch, *A. pectinatus* (Hook.) Dougl., is common in dry prairie on light and medium-textured soils. It has a deeply penetrating taproot, from which one to five straggling stems arise. The leaves are composed of 9–17 pairs of very narrow leaflets, and the large white or yellowish flowers are borne in short, dense heads. Two-grooved milk-vetch, *A. bisulcatus* (Hook.) A. Gray, is common on heavier-textured soils and in somewhat moister situations. Its deeply penetrating root gives rise to several erect stems. The leaves are composed of 13–29 pairs of elliptic leaflets, and the purplish flowers are pendant in elongated heads. This species has a distinctly unpleasant odor.

Both species are eaten only when overgrazing has reduced the available palatable forage, and death or severe illness can result.

In the Rocky Mountains timber milk-vetch, *A. miser* Dougl. ex Hook., and its variety *serotinus* (Gray) Barneby are poisonous to cattle, but poisoning is not the result of selenium accumulation in this case.

Spotted water-hemlock

Cicuta maculata L. var. *angustifolia* Hook.

Spotted water-hemlock, or water-hemlock, is considered the most poisonous plant of the north temperate zone. It is extremely toxic to all livestock, as well as to humans. Greatest toxicity occurs in the spring when the plants emerge.

Spotted water-hemlock is fairly common in wet places. It can grow up to 2 m tall, and its hollow stem can form several branches. The leaves are twice divided; the leaflets are lanceolate or linear-lanceolate, 5-8 cm long and 1-2 cm wide, with sharply toothed margins. The root system consists of a thickened lower part of the stem, to which one to several bulbous roots are attached. In cross section the roots are divided into several chambers that are separated by membranes. The cut surfaces exude a yellowish, oily substance that appears to contain the toxin cicutoxin.

A single root pulled out of the soft soil with the emerging foliage in spring can kill a mature cow when it is ingested. Poisoning in humans has occurred after sucking on roots and ingesting the sap.

Two other species, the bulbous water-hemlock, *C. bulbifera* L., and *C. mackenzieana* Raup, occur in the boreal forest zone in marshy areas and lake margins. Both are very poisonous.

Spotted water-hemlock is very similar in appearance to the nonpoisonous water-parsnip, *Sium suave* Walt., and the two plants may occur together in slough margins. They can be distinguished by their foliage, inflorescence, and roots. In spotted water-hemlock, the roots are hollow and chambered; the inflorescence is without bractlets; and the leaves are twice divided, each leaflet being lanceolate and growing up to 8 cm long. In water-parsnip the roots are solid; the inflorescence has bractlets; and the leaves are once divided, each leaflet being linear and growing up to 10 cm long.

A.C. Budd

Rushes and sedges

Baltic rush *Juncus balticus* Willd.

Baltic rush is so called because of its prevalence along the shores of the Baltic Sea. However, it is widespread, found throughout Europe, Asia, and North America. Dense stands seldom occur in the prairie area, but it is prevalent on all soils with a high water table, and sparse stands occur on nearly all sandy land. Because of the toughness of its stem, it is also known as wire rush.

Baltic rush has strongly developed, dark brown rootstocks, as well as an extensive and deeply penetrating fibrous root system. Numerous stems rise from the rootstocks, each terminating in a nodding flower cluster. A long bract or leaf continues from the stem above the flower head. Small, scaly leaves surround the stems at their base.

At sites where soil moisture is available throughout the summer, Baltic rush maintains a continuous growth. While it is green and in sparse stands mixed with grass, it is eaten readily; when cut early enough and properly cured, it makes a palatable but light component of native hay. However, it loses palatability when it dries out or after frosts. It has a higher protein content than many range grasses, but its fiber content is relatively low. Because its extensive root system is an efficient soil binder, it is an excellent plant for erosion control. Children of pioneer settlers used the wiry stems to make "daisy chains", and some Indian tribes used the stems to make baskets and mats.

Although Baltic rush is the most common of the *Juncus* group growing in the prairie area, 24 other species also occur. Toad rush, *J. bufonius* L., is common around sloughs and in moist areas throughout the Prairie Provinces. Several other species occur in meadows, occasionally in some abundance, and are eaten when young. None, however, is of great importance.

A.C. Budd

Creeping spike-rush

Eleocharis palustris (L.) Roem. & Schult.

Creeping spike-rush, or spike-rush, is a common member of the Cyperaceae family. It inhabits the muddy, moist shores of marshes and grows profusely in shallow sloughs throughout the southern portion of the prairie area. It also persists in areas of considerable salinity, provided its moisture requirements are satisfied. It has a wide distribution throughout Canada, the United States, and Eurasia.

This perennial plant grows from a creeping rootstock. Its numerous, tough roots fill the soil and send up tufts of stems at close intervals to heights of 35–50 cm. The stems are round or oval and are filled with pith. A few short leaves may develop around the base, but the rounded stem is the prominent feature of the plant. Each floret is complete, containing both male and female organs. The seeds are yellowish brown and shiny.

Creeping spike-rush is palatable to horses, so palatable in fact that they wade into sloughs to graze it during nearly all stages of its growth. Cattle eat it readily, but less avidly than horses do. During the spring and summer, it has high protein, phosphorus, and carbohydrate contents and is low in fiber and fat. It does not make good hay, because its palatability seems to disappear with the drying of the slough. It can be fed as an emergency winter roughage but should be supplemented with both grain and better-quality hay.

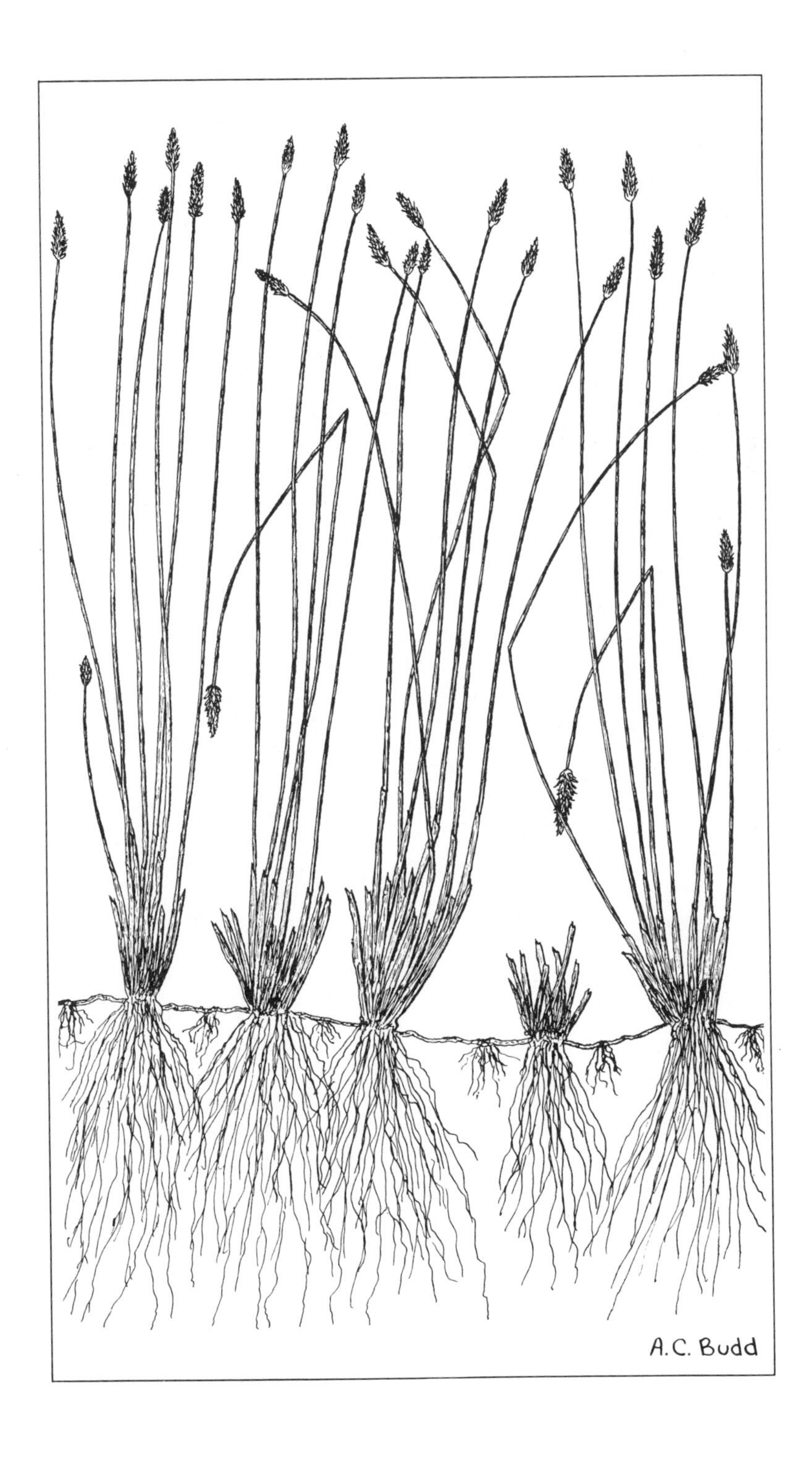
A.C. Budd

Awned sedge *Carex atherodes* Spreng.

Awned sedge is one of the most important plants growing in fresh to brackish marshes and shallow sloughs in the prairie area. It also has a wide distribution beyond the region, as it is found in similar sites in most of North America and throughout northern Eurasia.

Awned sedge grows from strong rootstocks. Numerous three-angled leafy stems arise from these at relatively close intervals; in fact, over 400 stems per square metre have been counted. Leaves are three-ranked and soft, with the underside covered by a dense mass of short hairs. Although the leaves hang down, when they are lifted to their full height their tips may be more than 1 m from the ground. The male flower head grows at the end of the stem and is dark brown and narrow. The female spike grows among the upper leaves; it is brown and thick when mature and each fruit has a long beak with two, often divergent teeth. Stands of the type described above yield up to 4 t of palatable, nutritious, aromatic hay per hectare. To obtain such yields the stands should be protected from grazing and enough water should be available to flood the stand at least until early July. The hay should be cut as soon as possible after the soil surface falls dry.

Awned sedge makes palatable pasture for cattle and sheep but does not tolerate heavy grazing. Also, it often occurs together with water sedge, *C. aquatilis* Wahlenb., recognizable by its bluish leaves and narrow female spikes that bear fruits without beaks. Water sedge is not palatable and its presence can be the cause of overgrazing other, more palatable plants.

A.C. Budd

Beaked sedge *Carex rostrata* Stokes

In the parklands and farther north, beaked sedge is as important for slough hay as awned sedge is in southern areas. The yield, protein content, and tolerance to haying or grazing are very similar in both species. Both species may occur in a mixed stand, with awned sedge dominant in the south and beaked sedge predominating in the north.

Beaked sedge grows from strong creeping rootstocks. Its soft leaves grow in whorls of three and are entirely free from hairs. The male spike grows near the tip of the robust, three-angled stem, and the thick female spike grows among the leaves. An important difference between awned sedge and beaked sedge occurs in the female spikes; namely, in beaked sedge the seed ends in a distinct curved beak. Other differences are the smooth sheaths and the strong crossveining of leaves and sheaths in beaked sedge.

Beaked sedge is fairly nutritious, with about 18% crude protein until midsummer and, when accessible, it is readily grazed by cattle. As hay it is of considerably lower quality because it can rarely be cut when it is at its best. Moreover, it usually grows together with several other sedges, marsh grasses, and rushes, most of which are of poor quality and often of doubtful palatability. Marsh or slough hay, therefore, should not be fed without a supplement of high-protein feed.

a.c. Budd.

Low sedge

Carex stenophylla Wahlenb. ssp. *eleocharis* (Bailey) Hult.

Low sedge is one of the most common dryland plants of the Interior Plains of North America. Its range extends from the Yukon to the Gulf of Mexico, and it may be found at elevations of 300–2200 m above sea level. It has been collected throughout the entire prairie area, with the heaviest stands occurring on overgrazed or eroded rangelands.

Low sedge is the smallest of the dryland sedges. A shallow rootstock sends up single plants at distances of 3–10 cm apart. These plants consist of a central three-angled seed stalk and either three or six leaves growing from the crown. The leaves are about 2 mm wide and seldom more than 5–8 cm long. The single seed head on each plant has male flowers at its tip and female flowers immediately below, and its total length is seldom more than 2 cm. The seed matures early in June, and the brown heads and drying leaves are distinctive at a season when grasses are green. Heavy stands of low sedge indicate an overgrazed pasture. A casual observer may confuse the brownish color in June with carryover of grasses. However, on closer observation one can differentiate between the brown color of maturing sedge and the gray brown of carryover.

Low sedge seems to be fairly palatable, but its low growth and poor yield and the fact that it spreads as overgrazing progresses make it little better than a pasture weed. Where it becomes dominant, stockmen are advised to cultivate the pasture and reseed with grasses recommended for the district.

Two other small dryland sedges are often confused with low sedge and both may occur with it. Sun-loving sedge, *C. pensylvanica* Lam., has wider, glossier, and more numerous leaves and it usually has more than one flower cluster on its seed-bearing stem. Blunt sedge, *C. obtusata* Liljeblad., is very similar to sun-loving sedge except that its seeds are nearly black when mature. Neither of these sedges produces much forage, although both are apparently palatable.

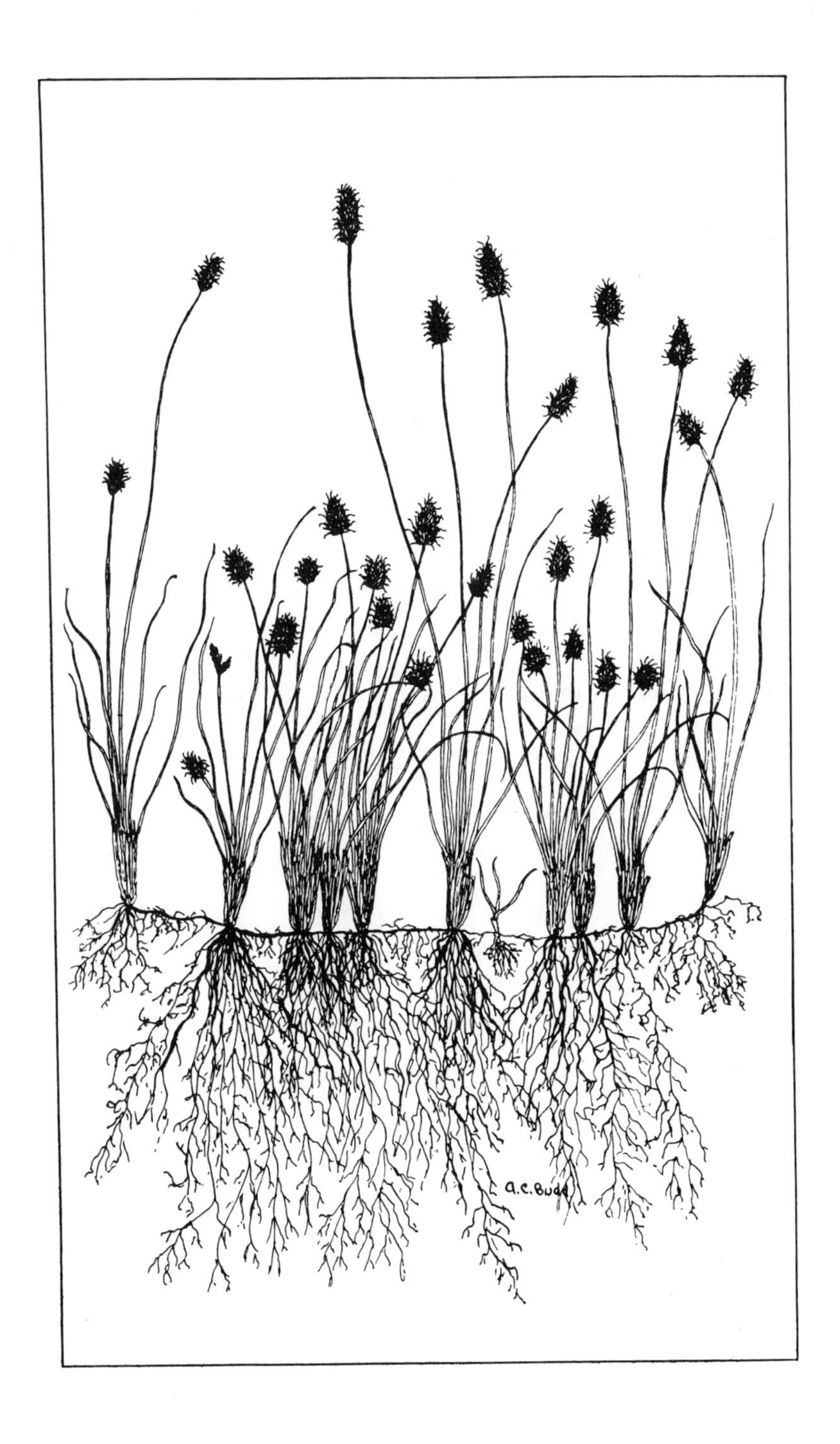
a.c.Budd

Thread-leaved sedge *Carex filifolia* Nutt.

Thread-leaved sedge, also known as niggerwool or hair sedge, is a short-growing, dryland member of the sedge family. Its distribution includes all of western North America from the Yukon to California and throughout the northern Interior Plains. It is more common in dry regions than where moisture is abundant. Its heaviest stands occur on sandy flats and exposed gravelly ridges, and at other sites where the soil may be shallow or infertile.

Thread-leaved sedge is easily recognized by its fine, threadlike, upright, deep green, rolled, glossy leaves, which may attain a length of 15 cm. Plants develop dense clumps, so dense that an observer would call the several small plants a single bunch grass; this appearance is caused by the numerous crowded shoots growing from the rootstock. Male and female flowers grow in the one head, with the male growing at the tip of the spike and the female at the base. The spike is oval-shaped, located at the extreme tip of a stem, and shiny after the seeds have ripened. A sure character for identifying thread-leaved sedge is the chestnut brown, dead leafage that surrounds the base of the leaves.

Thread-leaved sedge is one of the most common plants growing in the short-grass and mixed-grass regions of the prairie area. In a few districts it is dominant, but generally it is subdominant, occurring with blue grama, needle-and-thread, and June grass. It is usually more palatable than the associated grasses, and chemical analyses suggest that it contains higher protein and phosphorus and much lower crude fiber. Its numerous fine, tough, black roots fill the soil to a depth of 1 m and thus the species provides excellent protection against soil erosion. It withstands grazing better than the grasses with which it grows and, although trampling may tear the plant, this severe treatment seldom affects its future growth.

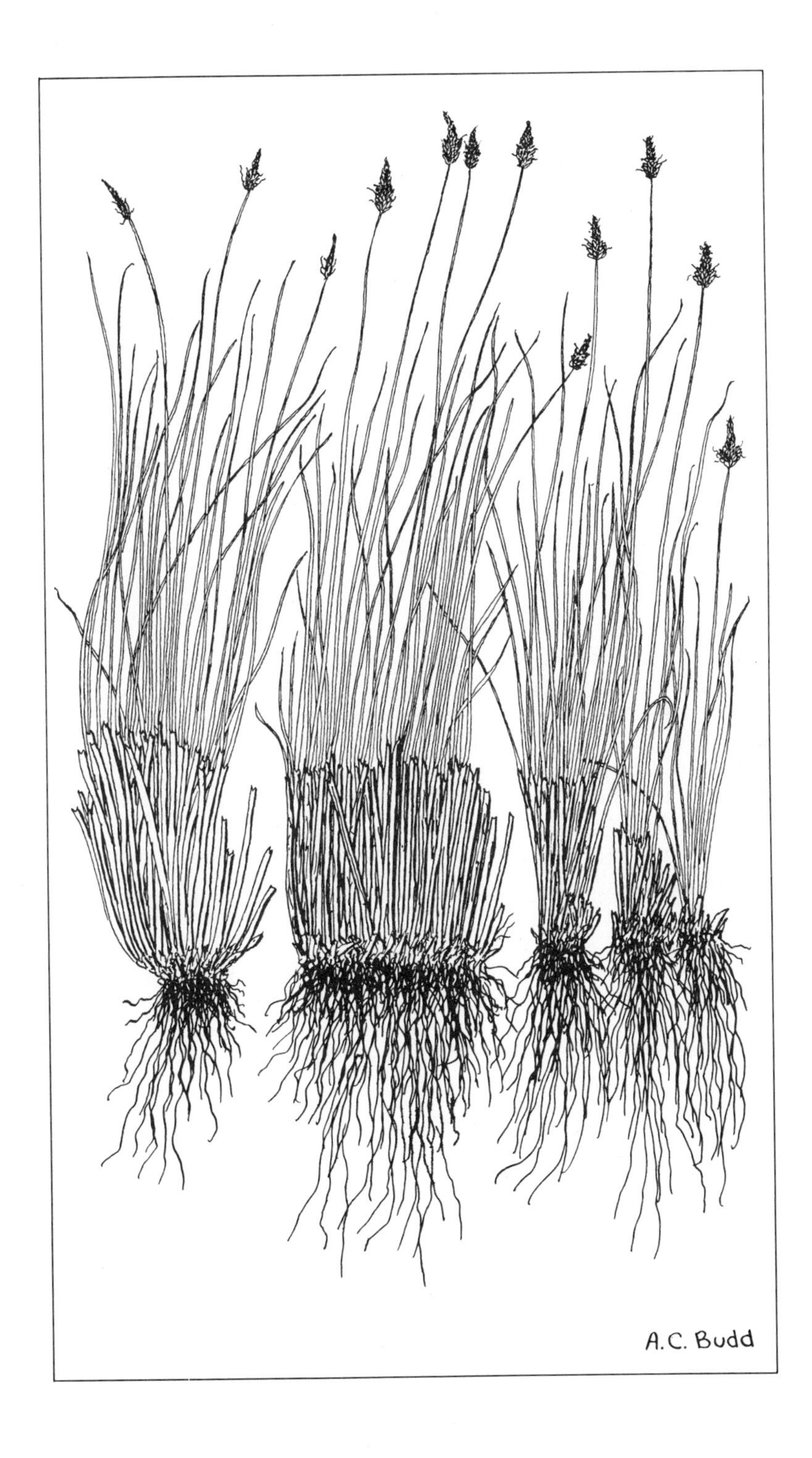
A.C. Budd

Trees and shrubs

Nuttall's atriplex *Atriplex nuttallii* S. Wats.

Nuttall's atriplex is a member of the goosefoot family. It occurs in the southern parts of Alberta and Saskatchewan and extends southward through Colorado and Nevada. It grows well on moderately saline soils in association with western wheat grass and on shallow soils dominated by blue grama.

Nuttall's atriplex is a low-growing, perennial shrub whose branches seldom attain a height of more than 70 cm. The leaves are usually alternate, long-oval, without stems, pale green, and scurfy. The yellowish male flowers grow on one plant and at the end of the branches, whereas the female flowers are on other plants and are usually surrounded with leaves. Its deep root system assures its protection against drought, although new stems grow slowly after being grazed or destroyed by dry conditions.

The palatability rating of Nuttall's atriplex is high for all classes of livestock despite its unappetizing appearance. Although it is grazed at any season, reports indicate that it is preferred in the autumn and winter. Chemical analyses show protein content of over 10% in late autumn, whereas phosphorus content is nearly twice as great as for the cured grasses. As with many other shrubs its crude fiber content is low, whereas that of ash is high.

Members of the *Atriplex* group are found throughout the temperate and tropical portions of the world. They are the dominant shrubs in the deserts of Australia and in many of the salt flats throughout the western United States. They are rated as good to excellent forage plants wherever they grow, but require careful management to maintain their productivity.

a.C.Budd

Winterfat *Eurotia lanata* (Pursh) Moq.

Winterfat, or white-sage, is a member of the goosefoot family whose range coincides closely with the semiarid area of western North America. Both its scientific names are derived from Latin words, *Eurotia* referring to the white, hairy herbage and *lanata* to the dense, woolly growth covering the leaves. It grows on all well-drained soils but may occur on shallow, eroded sites as well as on soils containing considerable alkali.

Winterfat is a low-growing shrub, seldom more than 50 cm tall. Numerous branches grow from the woody base, each of which produces many linear leaves. The male and female flowers are separate, with the male growing above the female. Its ash gray color and hairiness are its most distinguishing features during the summer.

The palatability rating of winterfat is very high for all domestic livestock and game. Its nutritive content is in good balance, with protein and phosphorus being exceptionally high even in late fall. The deeply penetrating root system assures a moisture and food supply at all times. Its rapid recovery after drought is a feature of its growth.

Unfortunately, winterfat can be grazed out and has been on many range areas in the western United States. The palatability of all parts of the plant and its susceptibility to damage by tramping explain why it disappears from the cover. However, it can be maintained in a productive condition when the associated grasses are managed so that they maintain a carryover of 40–50% during years of average growth. Overuse of winterfat for 1 or 2 years does not reduce its stand, provided a good growth season reoccurs to revitalize old plants and to produce seed. Winterfat is a good ornamental plant, easily grown under dry conditions.

A.C. Budd

Chokecherry *Prunus virginiana* L.

Chokecherry is a common shrub or small tree growing up to 6 m tall, occurring throughout the prairie area and in the parklands. It is particularly common on light soils and in sandhill areas, as well as in moist coulees and draws. It often occurs together with the saskatoon.

Chokecherry leaves are obovate or oval, 2–8 cm long, smooth, and with finely toothed margins. Flowers are numerous, about 1 cm across, occurring in dense, hanging racemes. The fruit is a small cherry that is red in the typical variety occurring in the southeastern part of the area, or black in the common western form, var. *melanocarpa* (A. Nels.) Sarg. In both areas a yellow-fruited form is occasionally encountered.

Both cattle and sheep browse on the chokecherry, especially on the emerging leaves in the spring and in the fall. Because the foliage develops prussic acid when injured by frost or drought, browsing animals can be poisoned by it.

Roses *Rosa* spp.

Roses have a widespread distribution throughout the temperate zone of North America where nearly 100 native species have been identified. In the prairie area four wild species are found, but only three are common: the prickly rose, *R. acicularis* Lindl., which grows on the open plains and around bluffs; the prairie rose, *R. arkansana* T. Porter; and Wood's rose, *R. woodsii* Lindl., which blooms at the margin of bluffs and in open woods on sandy soils.

Roses as a group are easy to distinguish from other plants. Their brown, spiny stems; small, green, and saw-edged leaves; white, pink, or red flowers; and red fruits are characters easily recognized. However, it is often difficult to identify one rose species from another, not only because the characters of a species are variable, but because the species hybridize readily to create plants having characters of two or more species.

Palatability of roses varies greatly among species, districts, and livestock. Sheep browse the leaves and fruits of nearly all species and in fact overgraze sparse or open stands. Cattle eat Wood's rose readily, especially in the autumn and on ranges where browse is more abundant than grass. In general, roses can be classified as less palatable than most grasses, but as palatable as most browse species, especially for sheep. Nutritive values, as determined by chemical analyses, are fairly well balanced, with a protein content of 10–12% during July and August. Phosphorus content is not so high as that of aspen but still well above that of associated grasses. The fiber content of the leaves is low.

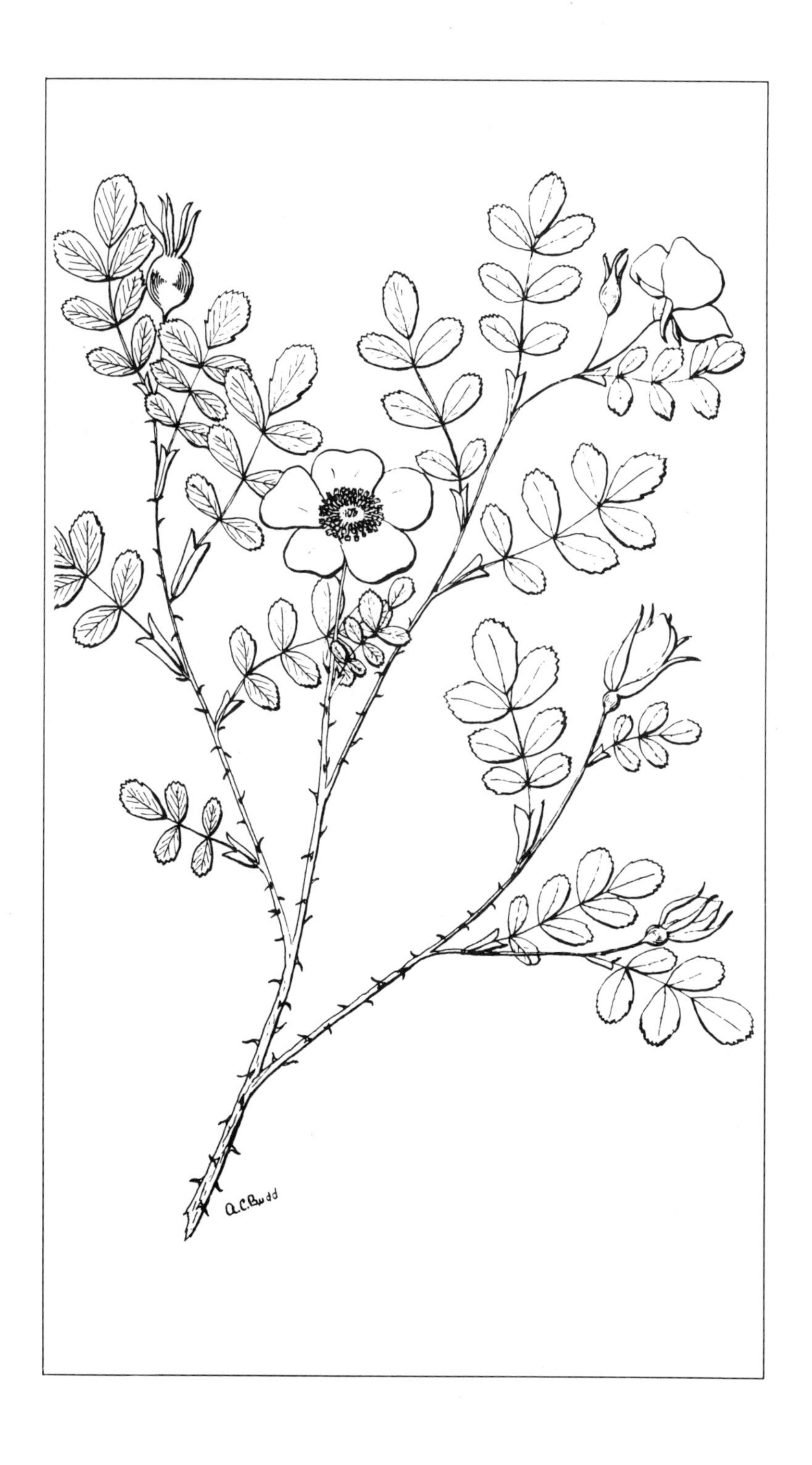
A.C.Budd

Saskatoon *Amelanchier alnifolia* Nutt.

Saskatoon, shadberry, serviceberry, and June-berry are all common names for this native member of the rose family. It is found throughout the temperate portions of North America and grows under a range of conditions from dry sunny slopes to shaded sites in coniferous forests. In the prairie area it is an inhabitant of shaded coulees and the adjacent upland.

Saskatoon may occur as either a dense shrub or a small tree, although the shrub type is the more common. The ash gray stems produce an abundance of round or round-oval leaves that are smooth on the upper side but hairy below. The upper portion of the leaf margin is saw toothed, but the lower third or half is entire; 10–14 pairs of conspicuous veins extend from the mid rib to the edge of the leaf. The round fruit is purplish with a bloom.

Because of its wide distribution, palatability, and availability to grazing animals, saskatoon is recognized as one of the best browse plants for all classes of domestic livestock and game. As with other browse plants there is greater utilization after midsummer. Chemical analyses indicate a high content of phosphorus during the entire year and high protein content. Grazing may reduce its vitality but saskatoon is one of the most persistent of the browse species. Livestock grazing the outer tips promote a greater growth of the inner foliage, which in time may develop into short trees. Thus moderate rates of grazing improve the stand of this species.

There are several species of *Amelanchier* growing in North America, although saskatoon is the only one common in the prairie area. All species produce edible fruit. Saskatoon foliage may cause poisoning in early spring when browsed, because the leaves contain prussic acid in early stages of growth.

A.C. Budd

Shrubby cinquefoil *Potentilla fruticosa* L.

A common range plant of the rose family is shrubby cinquefoil. It has a wide-spread distribution throughout Asia, Europe, and North America. It prefers valleys, bogs, or cool, moist plateaus. In the prairie area it is abundant in the Foothills and Cypress Hills, but small stunted stands are found throughout the region. The name *fruticosa* refers to the shrubby type of growth. Other common names are yellow rose and buck brush.

Shrubby cinquefoil grows from a woody crown that shows annular growth rings. The numerous stems 1–1.5 m tall are covered with shreddy, thin bark, and the young branches are clothed with grayish hairs. The leaves have three to seven leaflets, whose bases are so close together that the leaf might be considered to be palmate. The leaflets are leathery, hairy, and pointed at each end. The plant produces yellow flowers and blooms from June until freeze-up.

Shrubby cinquefoil is seldom eaten as long as other food is available. When sparse stands are heavily grazed or when heavy stands increase in density and size, overutilization of the associated palatable grasses is likely to occur. Sheep utilize cinquefoil more extensively than do other domestic livestock or game. The surest method of control is to mow dense stands every 3rd or 4th year, but rotations and moderate grazing rates do serve to maintain associated grasses and forbs. Shrubby cinquefoil is easily grown and is valued as an ornamental because it produces a profusion of flowers during a long season. As with many other browse plants, cinquefoil leaves have a relatively high protein content, namely 12–15% at the end of July, a low fiber content, and considerably more phosphorus than the associated grasses.

A.C. Budd

Silver sagebrush *Artemisia cana* Pursh

Silver sagebrush is a shrub member of the thistle family that inhabits the central and western portions of the prairie area. It is found northward as far as the parkland and extends southward through Montana and Utah. Silver sagebrush belongs to the dry plains and sand region and produces its heaviest-known stands in the Great Sand Hills of western Saskatchewan.

The shrub grows 0.5–1.5 m tall and has a deeply penetrating tap root. Its bark is gray and often shredded. The leaves are up to 40 mm long, silvery hairy on both sides, and occasionally toothed at the end. The tiny, yellow florets are crowded into a leafy panicle.

A closely related species, known as big sagebrush or black-sage, *A. tridentata* Nutt., occurs only at one known point in the Rocky Mountains in Alberta. However, it is the dominant species over extensive areas in central British Columbia and throughout the intermountain area of the United States. It is distinguished from silver sagebrush by the three indentations at the tips of the leaves, as well as by its larger size.

Silver sagebrush cannot be classified as palatable, although it is eaten on occasion. Sparse stands on sheep range disappear quickly, but heavy stands seldom show any evidence of utilization. Cattle do not relish the plant but thrive on it when forced to graze it. Pronghorn antelope use sagebrush flats as winter range throughout the plains. Chemical analyses show that the leaves of silver sagebrush have about 20% protein in summer and maintain 10% protein in fall. Phosphorus and fat content also remain high, up to twice the level of that in associated grasses.

a.C.Budd.

Water birch *Betula occidentalis* Hook.

Water birch is the most common birch growing in the prairie area. It occurs on moist flats along river banks and in moist, sandy flats from Manitoba to the west coast and from the Yukon to California. In the prairie area its largest stands are found in the Great Sand Hills in western Saskatchewan. It has many common local names including river birch, mountain birch, swamp birch, and black birch.

Water birch occurs either as a short tree or as a tall, dense shrub. Its dark brown, aromatic, resinous bark is dotted with gray lenticels. The leaves are seldom over 5 cm long and the edges have a double serration, that is, the teeth appear to be in pairs with one larger than the other. As with other birch species, it has separate male and female flowers, but both grow on the same tree. The fruit is a small nut enclosed by a scale in a small cone.

Birches generally are not good browse; in fact they are eaten only by very hungry animals. However, ranchers from the Great Sand Hills report that cattle do break off the branches to reach the young stems during the winter and early spring.

There are a few other species of birch occurring within the prairie area. Swamp birch, *B. glandulifera* (Regel) Butler, is quite common in acid soils in mountain and northern swamps, and paper birch, *B. papyrifera* Marsh., is common in the boreal forest region on sandy soils. A closely related group, the alders, resembles the birches and may be confused with them. Characters of the alders that help to separate them from the birches include the smoother bark and the rounder leaves. Birches and alders are poor forage plants.

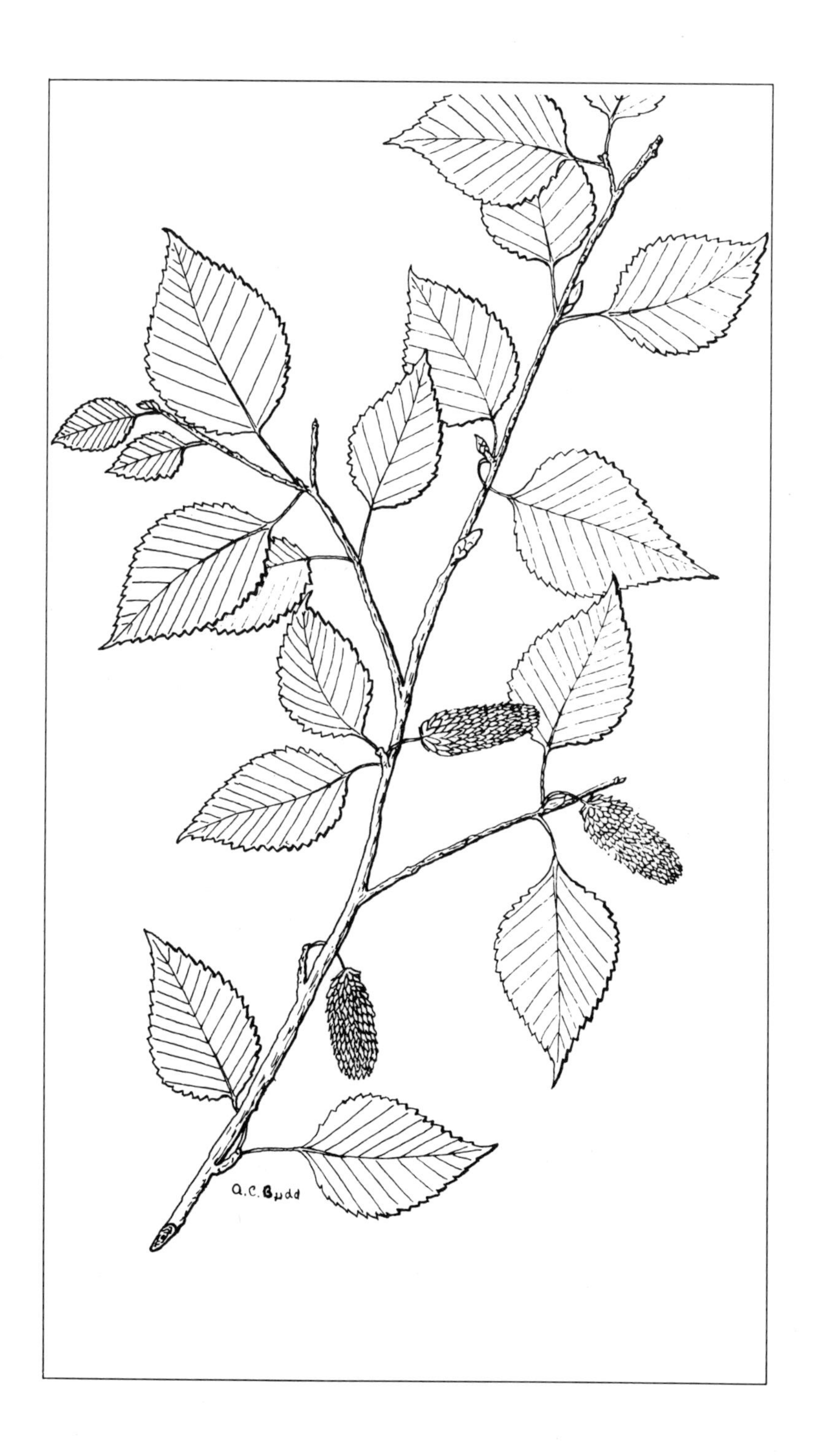
A.C. Budd

Aspen poplar *Populus tremuloides* Michx.

Aspen poplar is the most common tree throughout the prairie area. It has a wide range, occurring from Newfoundland to Alaska and southward as fas as Tennessee in the east and New Mexico in the west. Since the early 1900s it has increased its stand in the parkland and has spread southward into the drier portions of Alberta and Saskatchewan. It has many local names including trembling or quaking poplar, trembling aspen, and popple, all referring to the trembling motion of the leaves in very light breezes.

Aspen poplar is a forest-forming tree, although it grows also in open stands. Its smooth, whitish or greenish bark covers most of the trunk, but rough, black bark may cover the first 2 m, and black spots mark the bark at higher elevations, becoming smaller and fewer upward. The leaf stalks are flattened and bear smooth, toothed leaves 3–8 cm wide. The male and female flowers, which occur on different trees, appear before the leaves in the form of pendulous catkins. This species reproduces strongly by root shoots, but heavy grazing tends to slow this development.

Nearly all livestock including game graze aspen poplar during all seasons of the year, but seldom heavily enough to check its spread. Leaves as well as young twigs are eaten readily. Chemical analyses of this species indicate high protein and phosphorus and extremely low crude fiber contents, all characteristics of good forage.

As aspen poplar grows on nearly all soil types where climate is favorable, its presence is not a good indicator of soil quality. However, on poor soil its growth is slow and spindly, and its associates are range weeds. On fertile soils it develops open stands, with individual trees up to 20 m tall; where these conditions occur aspen poplar grows in association with palatable and nutritious grasses and forbs, a combination that makes excellent pasturage.

A.C.Budd

Willows *Salix* spp.

The willows contain many species, most of which are difficult to identify. They have a wide range, also, growing on moist slopes and flats, along river banks, and in mountain meadows throughout the northern hemisphere. They may be very small, slender shrubs or large trees suitable for lumber. Studies of the willows growing in the prairie area indicate that there are over 30 species, all shrubs or small, gnarled trees.

The willows and poplars have many characters in common and in some instances are hard to distinguish. However, in the prairie area poplars occur as trees, whereas willows grow as shrubs. Poplars usually have a long leaf stalk whereas willow leaves have short or very short stalks. The buds of poplars are covered with several sticky scales, whereas the buds of willows have a single shiny scale whose color is usually dark brown. Poplar leaves usually have saw-toothed edges, whereas willow leaves have smooth or only finely serrated edges. As with the poplar, the willows have male and female flowers on separate trees and in most cases the catkins appear before the leaves. A few species have creeping rootstocks.

The willows make good browse. Sheep and deer eat the leaves and twigs more readily than do cattle, but cattle probably utilize them more fully because of their grazing habits. Chemical analyses show that the willows have higher protein and phosphorus contents than do the grasses, and the young growth is lower in crude fiber.

Because willows grow only in moist locations, the species do not spread into dry grassland as does the aspen poplar. In fact, because heavy grazing reduces the grass cover, it usually also reduces the willow population. Thus the disappearance of willows, particularly the broad-leaved shrubby types, may be an indication that overgrazing is occurring.

A.C. Budd

KEY TO PREFERRED COMMON NAMES

Species descriptions for the range and forage plants in this publication appear in alphabetic order under the main headings and subheadings by which they are organized. The alphabetic listings of the main headings, the subheadings, and the species descriptions are ordered according to the preferred common names. The organization used in this publication is given in the Contents. For easy reference, the main headings are repeated in the runningheads on the left-hand pages, with the corresponding subheadings appearing in the runningheads on the right-hand pages. The common name under which a species is described is used as the subheading whenever only one species in the group is treated.

In this key the left column contains the preferred common names for each species described in the text, in boldface roman type; the scientific names for each of these species, in lightface italic type; and the alternate common and scientific names that are mentioned in the text as being used regionally for these species, in lightface roman type. The corresponding entries in the right column give the main heading under which the species description appears; the appropriate subheading whenever it differs from the species name; and, for each scientific and alternate common or scientific name, the preferred common name under which the species description can be found in the text in boldface type.

To locate the species description in the text, either consult the Contents for the exact page number under the main heading and subheading given in the key, or simply flip through the pages, scanning the runningheads, until you locate the appropriate main heading and subheading and turn there to the description of the species you are looking for.

The key also includes in the left column all common and scientific names of species that are mentioned in the main descriptions of other species but which are not described separately, with reference to the species description that contains their mention in the right column, as described above.

To find...	look under...
Agropyron cristatum	grasses, wheat grasses: **crested wheat grass**
Agropyron dasystachyum	grasses, wheat grasses: **northern wheat grass**
Agropyron elongatum	grasses, wheat grasses: **tall wheat grass**
Agropyron intermedium	grasses, wheat grasses: **intermediate wheat grass**
Agropyron repens	grasses, wheat grasses: **quack grass**
Agropyron riparium	grasses, wheat grasses: **streambank wheat grass**
Agropyron smithii	grasses, wheat grasses: **western wheat grass**
Agropyron trachycaulum var. *trachycaulum*	grasses, wheat grasses: **slender wheat grass**
Agropyron trachycaulum var. *unilaterale*	grasses, wheat grasses: **awned wheat grass**
Agrostis scabra	grasses, bent grasses: **rough hair grass**
Agrostis stolonifera	grasses, bent grasses: **creeping bent grass**
alders	*see* trees and shrubs: **water birch**

To find . . .	look under . . .
alexander	*see* herbs, parsleys: **snakeroot**
alfalfa	legumes
alkali cord grass	grasses, cord grasses
alkali grass	grasses: **desert salt grass**
alpine timothy	grasses, timothys
alsike clover	*see* legumes: **clovers**
Altai wild rye grass	grasses, wild rye grasses
Amelanchier alnifolia	trees and shrubs, rose family: **saskatoon**
American hedysarum	legumes
American milk-vetch	*see* legumes, milk-vetches: **ground-plum**
American vetch	*see* legumes: **vetches**
Andropogon gerardi	grasses, bluestems: **big bluestem**
Andropogon scoparius	grasses, bluestems: **little bluestem**
arrow-grasses	poisonous plants
Artemisia cana	trees and shrubs: **silver sagebrush**
Artemisia frigida	herbs: **pasture sage**
Artemisia tridentata	*see* trees and shrubs: **silver sagebrush**
ascending purple milk-vetch	*see* legumes, milk-vetches: **ground-plum**
aspen poplar	trees and shrubs, willow family
asters	herbs, thistle family
Aster spp.	herbs, thistle family: **asters**
Aster conspicuus	*see* herbs, thistle family: **asters**
Aster laevis	*see* herbs, thistle family: **asters**
Astragalus spp.	poisonous plants: **milk-vetches.** *See also* poisonous plants: **locoweeds**
Astragalus bisulcatus	*see* poisonous plants: **milk-vetches**
Astragalus cicer	legumes, milk-vetches: **cicer milk-vetch**
Astragalus crassicarpus	legumes, milk-vetches: **ground-plum**
Astragalus danicus	*see* legumes, milk-vetches: **ground-plum**
Astragalus frigidus	*see* legumes, milk-vetches: **ground-plum**
Astragalus miser	*see* poisonous plants: **milk-vetches**
Astragalus miser var. *serotinus*	*see* poisonous plants: **milk-vetches**
Astragalus pectinatus	*see* poisonous plants: **milk-vetches**
Astragalus striatus	*see* legumes, milk-vetches: **ground-plum**
Atriplex nuttallii	trees and shrubs, goosefoots: **Nuttall's atriplex**
awned sedge	rushes and sedges, sedges
awned wheat grass	grasses, wheat grasses
awnless bromegrass	grasses, brome grasses: **smooth brome**
Baltic rush	rushes and sedges, rushes
beaked sedge	rushes and sedges, sedges
Beckmannia syzigachne	grasses: **slough grass**
bent grasses	grasses
Bessey's locoweed	*see* poisonous plants: **locoweeds**
Betula glandulifera	*see* trees and shrubs: **water birch**
Betula occidentalis	trees and shrubs: **water birch**
Betula papyrifera	*see* trees and shrubs: **water birch**
big bluestem	grasses, bluestems
big sagebrush	*see* trees and shrubs: **silver sagebrush**
bird's-foot trefoil	legumes
biscuitroot	*see* herbs, parsleys: **snakeroot**

To find...	look under...
black birch	trees and shrubs: **water birch**
black-sage	*see* trees and shrubs: **silver sagebrush**
bluebunch fescue	grasses, fescues
blue grama	grasses
blue grasses	grasses
blue-joint	grasses, reed grasses
bluejoint	grasses, wheat grasses: **western wheat grass**
bluejoint turkeyfoot	grasses, bluestems: **big bluestem**
bluestem	grasses, wheat grasses: **western wheat grass**
bluestems	grasses
blunt-fruited sweet cicely	*see* herbs, parsleys: **sweet cicelys**
blunt sedge	*see* rushes and sedges, sedges: **low sedge**
Bouteloua curtipendula	*see* grasses: **blue grama**
Bouteloua gracilis	grasses: **blue grama**
brome grasses	grasses
Bromus biebersteinii	grasses, brome grasses: **meadow brome grass**
Bromus ciliatus	grasses, brome grasses: **fringed brome grass**
Bromus inermis	grasses, brome grasses: **smooth brome**
Bromus tectorum	*see* grasses, brome grasses: **fringed brome grass**
broomsedge	grasses, bluestems: **little bluestem**
brown beard grass	grasses, bluestems: **little bluestem**
buck brush	trees and shrubs, rose family: **shrubby cinquefoil**
buckwheats	*see* herbs: **smartweeds**
bulbous water-hemlock	*see* poisonous plants: **spotted water-hemlock**
butterweeds	herbs, thistle family: **groundsels**
Calamagrostis canadensis	grasses, reed grasses: **blue-joint**
Calamagrostis inexpansa	grasses, reed grasses: **northern reed grass**
Calamagrostis montanensis	grasses, reed grasses: **plains reed grass**
Calamagrostis rubescens	grasses, reed grasses: **pine grass**
Calamovilfa longifolia	grasses: **sand grass**
Canada blue grass	grasses, blue grasses
Canada wild rye	grasses, wild rye grasses
Canby blue grass	grasses, blue grasses
Carex aquatilis	*see* rushes and sedges, sedges: **awned sedge**
Carex atherodes	rushes and sedges, sedges: **awned sedge**
Carex filifolia	rushes and sedges, sedges: **thread-leaved sedge**
Carex obtusata	*see* rushes and sedges, sedges: **low sedge**
Carex pensylvanica	*see* rushes and sedges, sedges: **low sedge**
Carex rostrata	rushes and sedges, sedges: **beaked sedge**
Carex stenophylla ssp. *eleocharis*	rushes and sedges, sedges: **low sedge**
Castilleja mineata	herbs: **red Indian paintbrush**
Chewing's fescue	*see* grasses, fescues: **creeping red fescue**
chokecherry	trees and shrubs, rose family
cicer milk-vetch	legumes, milk-vetches
Cicuta bulbifera	*see* poisonous plants: **spotted water-hemlock**
Cicuta mackenzieana	*see* poisonous plants: **spotted water-hemlock**
Cicuta maculata var. *angustifolia*	poisonous plants. *See also* herbs, parsleys: **sweet cicelys**
clovers	legumes
cock's-foot	grasses: **orchard grass**

To find...	look under...
Colorado bluestem	grasses, wheat grasses: **western wheat grass**
common hawkweed	herbs, thistle family: **umbellate hawkweed**
common spear grass	grasses, needle grasses: **needle-and-thread**
composites	*see* herbs: **pasture sage**
cord grasses	grasses
couch grass	grasses, wheat grasses: **quack grass**
cow-parsnip	*see* herbs, parsleys: **snakeroot**
creeping bent grass	grasses, bent grasses
creeping red fescue	grasses, fescues
creeping spike-rush	rushes and sedges, rushes
Crepis spp.	herbs, thistle family: **hawk's-beards**
crested wheat grass	grasses, wheat grasses
Cyperaceae	*see* rushes and sedges, rushes: **creeping spike rush**
Dactylis glomerata	grasses: **orchard grass**
dandelion	herbs, thistle family
Danthonia intermedia	grasses, oat grasses: **timber oat grass**
Danthonia parryi	grasses, oat grasses: **Parry oat grass**
death camas	poisonous plants
Delphinium spp.	poisonous plants: **larkspurs**
Delphinium bicolor	*see* poisonous plants: **larkspurs**
Delphinium glaucum	*see* poisonous plants: **larkspurs**
Deschampsia caespitosa	grasses: **tufted hair grass**
desert salt grass	grasses
Distichlis stricta	grasses: **desert salt grass**
downy brome	*see* grasses, brome grasses: **fringed brome grass**
downy wheat grass	grasses, wheat grasses: **northern wheat grass**
dropseeds	*see* grasses: **sand dropseed**
early blue grass	grasses, blue grasses
early yellow locoweed	*see* poisonous plants: **locoweeds**
Eleocharis palustris	rushes and sedges, rushes: **creeping spike-rush**
Elymus angustus	grasses, wild rye grasses: **Altai wild rye grass**
Elymus canadensis	grasses, wild rye grasses: **Canada wild rye**
Elymus innovatus	grasses, wild rye grasses: **hairy wild rye grass**
Elymus junceus	grasses, wild rye grasses: **Russian wild rye grass**
Elymus virginicus	grasses, wild rye grasses: **Virginia wild rye grass**
Epilobium angustifolium	herbs: **fireweed**
Eurotia lanata	trees and shrubs, goosefoots: **winterfat**
evening primrose	*see* herbs: **fireweed**
false melic	grasses, oat grasses: **purple oat grass**
false Solomon's-seal	herbs
fescues	grasses
Festuca campestris	grasses, fescues: **rough fescue**
Festuca elatior	*see* grasses, fescues: **tall fescue**
Festuca elatior var. *arundinacea*	grasses, fescues: **tall fescue**
Festuca fallax	*see* grasses, fescues: **creeping red fescue**
Festuca hallii	grasses, fescues: **plains rough fescue**
Festuca idahoensis	grasses, fescues: **bluebunch fescue**
Festuca ovina	*see* grasses, fescues: **bluebunch fescue**
Festuca rubra	grasses, fescues: **creeping red fescue**
Festuca rubra var. *commutata*	*see* grasses, fescues: **creeping red fescue**

To find...	look under...
figworts	*see* herbs: **red Indian paintbrush**
fireweed	herbs
flatstem blue grass	grasses, blue grasses: **Canada blue grass**
fowl manna grass	*see* grasses: **tall manna grass**
foxtail barley	grasses
fringed brome grass	grasses, brome grasses
fringed sage	herbs: **pasture sage**
Glyceria borealis	*see* grasses: **tall manna grass**
Glyceria grandis	grasses: **tall manna grass**
Glyceria striata	*see* grasses: **tall manna grass**
goosefoots	trees and shrubs
gramma grasses	*see* grasses: **blue gramma**
green feather grass	grasses, needle grasses: **green needle grass**
green needle grass	grasses, needle grasses
green stipa grass	*see* grasses, needle grasses: **green needle grass**
ground-plum	legumes, milk-vetches
groundsels	herbs, thistle family
hair grasses	*see* grasses: **tufted hair grass**
hair sedge	rushes and sedges, sedges: **thread-leaved sedge**
hairy wild rye grass	grasses, wild rye grasses
hawk's-beards	herbs, thistle family
Hedysarum alpinum var. *americanum*	legumes: **American hedysarum**
Helictotrichon hookeri	grasses, oat grasses: **Hooker's oat grass**
Heracleum lanatum	*see* herbs, parsleys: **snakeroot**
Hieracium umbellatum	herbs, thistle family: **umbellate hawkweed**
Hierochloe odorata	grasses: **sweet grass**
Hooker's oat grass	grasses, oat grasses
Hordeum jubatum	grasses: **foxtail barley**
Idaho fescue	grasses, fescues: **bluebunch fescue**
Indian grass	grasses
Indian paintbrush	herbs: **red Indian paintbrush**
Indian rice grass	grasses
intermediate wheat grass	grasses, wheat grasses
Juncus balticus	rushes and sedges, rushes: **Baltic rush**
Juncus bufonius	*see* rushes and sedges, rushes: **Baltic rush**
June-berry	trees and shrubs, rose family: **saskatoon**
June grass	grasses
Kentucky blue grass	grasses, blue grasses
Koeleria cristata	grasses: **June grass**
Koeleria gracilis	grasses: **June grass**
Koeleria macrantha	grasses: **June grass**
lady's-thumb	herbs: **smartweeds**
larkspurs	poisonous plants
late yellow locoweed	*see* poisonous plants: **locoweeds**
Lathyrus ochroleucus	*see* legumes: **purple vetchling**
Lathyrus venosus	legumes: **purple vetchling**
lilies	*see* poisonous plants: **death camas**
lily-of-the-valley	*see* herbs: **false Solomon's-seal**
little bluestem	grasses, bluestems
locoweeds	poisonous plants

To find...	look under...
Lomatium spp.	*see* herbs, parsleys: **snakeroot**
Lotus corniculatus	legumes: **bird's-foot trefoil**
Lotus pedunculatus	*see* legumes: **bird's-foot trefoil**
low larkspur	*see* poisonous plants: **larkspurs**
low sedge	rushes and sedges, sedges
lucerne	legumes: **alfalfa**
Lupinus argenteus	legumes: **silvery lupins**
Mandan wild ryegrass	*see* grasses, wild rye grasses: **Canada wild rye**
manna grasses	*see* grasses: **tall manna grass**
marsh arrow-grass	*see* poisonous plants: **arrow-grasses**
marsh ragwort	*see* herbs, thistle family: **groundsels**
marsh reed grass	grasses, reed grasses: **blue-joint**
mat muhly	grasses
meadow brome grass	grasses, brome grasses
meadow fescue	*see* grasses, fescues: **tall fescue**
Medicago sativa	legumes: **alfalfa**
Medicago sativa ssp. *falcata*	*see* legumes: **alfalfa**
Melilotus spp.	legumes: **sweet-clovers**
Melilotus alba	*see* legumes: **sweet-clovers**
Melilotus officinalis	*see* legumes: **sweet-clovers**
milk-vetches	legumes. *Also* poisonous plants
mountain birch	trees and shrubs: **water birch**
Muhlenbergia asperifolia	*see* grasses: **mat muhly**
Muhlenbergia richardsonis	grasses: **mat muhly**
muhly	*see* grasses: **mat muhly**
musineon	*see* herbs, parsleys: **snakeroot**
Musineon spp.	*see* herbs, parsleys: **snakeroot**
narrow-leaved American vetch	*see* legumes: **vetches**
narrow-leaved milk-vetch	*see* poisonous plants: **milk-vetches**
needle-and-thread	grasses, needle grasses
needle grasses	grasses
niggerwool	rushes and sedges, sedges: **thread-leaved sedge**
nodding wild rye grass	grasses, wild rye grasses: **Canada wild rye**
northern beard grass	grasses, bluestems: **big bluestem**
northern manna grass	*see* grasses: **tall manna grass**
northern porcupine grass	grasses, needle grasses
northern reed grass	grasses, reed grasses
northern wheat grass	grasses, wheat grasses
Nuttall's atriplex	trees and shrubs, goosefoots
Nuttall's salt-meadow grass	grasses
oat grasses	grasses
Onobrychis viciifolia	legumes: **sainfoin**
orchard grass	grasses
Oryzopsis hymenoides	grasses: **Indian rice grass**
Osmorhiza spp.	herbs, parsleys: **sweet cicelys**
Osmorhiza aristata	*see* herbs, parsleys: **sweet cicelys**
Osmorhiza chilensis	*see* herbs, parsleys: **sweet cicelys**
Oxytropis spp.	poisonous plants: **locoweeds**
Oxytropis besseyi	*see* poisonous plants: **locoweeds**
Oxytropis campestris ssp. *gracilis*	*see* poisonous plants: **locoweeds**

To find...	look under...
Oxytropis lambertii	*see* poisonous plants: **locoweeds**
Oxytropis sericea var. *spicata*	*see* poisonous plants: **locoweeds**
paint brushes	*see* herbs: **red Indian paintbrush**
panic grasses	*see* grasses: **switch grass**
Panicum virgatum	grasses: **switch grass**
paper birch	*see* trees and shrubs: **water birch**
Parry oat grass	grasses, oat grasses
parsleys	herbs
pasture sage	herbs
peavine	legumes: **purple vetchling**
perennial sow-thistle	herbs, thistle family
Perideridia spp.	*see* herbs, parsleys: **snakeroot**
persicaria	herbs: **smartweeds**
Phalaris arundinacea	grasses: **reed canary grass**
Phleum alpinum	grasses, timothys: **alpine timothy**
Phleum pratense	grasses, timothys: **timothy**
pine grass	grasses, reed grasses
pine reed grass	grasses, reed grasses: **pine grass**
plains reed grass	grasses, reed grasses
plains rough fescue	grasses, fescues
Poa canbyi	grasses, blue grasses: **Canby blue grass**
Poa compressa	grasses, blue grasses: **Canada blue grass**
Poa cusickii	grasses, blue grasses: **early blue grass**
Poa pratensis	grasses, blue grasses: **Kentucky blue grass**
Poa sandbergii	grasses, blue grasses: **Sandberg's blue grass**
Polygonum spp.	herbs: **smartweeds**
Polygonum amphibium	*see* herbs: **smartweeds**
popple	trees and shrubs, willow family: **aspen poplar**
Populus tremuloides	trees and shrubs, willow family: **aspen poplar**
porcupine grass	grasses, needle grasses
Potentilla fruticosa	trees and shrubs, rose family: **shrubby cinquefoil**
prairie cord grass	grasses, cord grasses
prairie beard grass	grasses, bluestems: **little bluestem**
prairie parsley	*see* herbs, parsleys: **snakeroot**
prairie rose	*see* trees and shrubs, rose family: **roses**
prickly rose	*see* trees and shrubs, rose family: **roses**
Prunus virginiana	trees and shrubs, rose family: **chokecherry**
Prunus virginiana var. *melanocarpa*	*see* trees and shrubs, rose family: **chokecherry**
Puccinellia nuttalliana	grasses: **Nuttall's salt-meadow grass**
purple locoweed	*see* poisonous plants: **locoweeds**
purple milk-vetch	*see* legumes, milk-vetches: **ground-plum**
purple oat grass	grasses, oat grasses
purple vetchling	legumes
quack grass	grasses, wheat grasses
quaking poplar	trees and shrubs, willow family: **aspen poplar**
ragworts	herbs, thistle family: **groundsels**
red clover	*see* legumes: **clovers**
red Indian paintbrush	herbs
red-seeded dandelion	*see* herbs, thistle family: **dandelion**

To find...	look under...
snakeroot	herbs, parsleys
Sonchus arvensis	herbs, thistle family: **perennial sow-thistle**
Sorghastrum nutans	grasses: **Indian grass**
sow-thistle	herbs, thistle family: **perennial sow-thistle**
spangletop	grasses
Spartina gracilis	grasses, cord grasses: **alkali cord grass**
Spartina pectinata	grasses, cord grasses: **prairie cord grass**
spike-rush	rushes and sedges, rushes: **creeping spike-rush**
Sporobolus cryptandrus	grasses: **sand dropseed**
spotted water-hemlock	poisonous plants. *See also* herbs, parsleys: **sweet cicelys**
squawroot	*see* herbs, parsleys: **snakeroot**
star-flowered Solomon's-seal	herbs: **false Solomon's-seal**
Stipa columbiana	*see* grasses, needle grasses: **green needle grass**
Stipa comata	grasses, needle grasses: **needle-and-thread**
Stipa curtiseta	grasses, needle grasses: **northern porcupine grass**
Stipa spartea	grasses, needle grasses: **porcupine grass**
Stipa viridula	grasses, needle grasses: **green needle grass**
streambank wheat grass	grasses, wheat grasses
sun-loving sedge	*see* rushes and sedges, sedges: **low sedge**
swamp birch	*see* trees and shrubs: **water birch**
sweet-anise	herbs, parsleys: **sweet cicelys**
sweet brome	legumes: **American hedysarum**
sweet cicelys	herbs, parsleys
sweet-clovers	legumes
sweet grass	grasses
sweetroot	herbs, parsleys: **sweet cicelys**
sweet vetch	legumes: **American hedysarum**
switch grass	grasses
tall fescue	grasses, fescues
tall larkspur	*see* poisonous plants: **larkspurs**
tall manna grass	grasses
tall wheat grass	grasses, wheat grasses
tansy ragwort	*see* herbs, thistle family: **groundsels**
Taraxacum erythrospermum	*see* herbs, thistle family: **dandelion**
Taraxacum officinale	herbs, thistle family: **dandelion**
thickspike wheat grass	grasses, wheat grasses: **northern wheat grass**
thistle family	herbs
thread-leaved sedge	rushes and sedges, sedges
three-flowered Solomon's-seal	herbs: **false Solomon's-seal**
tickle grass	grasses, bent grasses: **rough hair grass**
timber milk-vetch	*see* poisonous plants: **milk-vetches**
timber oat grass	grasses, oat grasses
timothy	grasses, timothys
timothys	grasses
toad rush	*see* rushes and sedges, rushes: **Baltic rush**
trembling aspen	trees and shrubs, willow family: **aspen poplar**
trembling poplar	trees and shrubs, willow family: **aspen poplar**
Trifolium spp.	legumes: **clovers**
Trifolium hybridum	*see* legumes: **clovers**
Trifolium pratense	*see* legumes: **clovers**